Arboviruses in Arthropod Cells In Vitro

Volume II

Editor

Conrad E. Yunker, Ph.D.

Cell Biologist
Hemoparasitic Diseases Research Unit
Agricultural Research Service
United States Department of Agriculture
and
Professor
Department of Veterinary Microbiology and Pathology
Washington State University
Pullman, Washington

CRC Press, Inc.
Boca Raton, Florida

Library of Congress Cataloging-in-Publication Data

Arboviruses in arthropod cells in vitro.

Includes bibliographies and index.
1. Arthropod-borne viruses. 2. Arthropod vectors.
3. Virus-vector relationships. 4. Cell culture.
I. Yunker, Conrad E. [DNLM: 1. Arboviruses--isolated &
purification. 2. Cells, Cultured. 3. Virus
Replication. QW 168.5.A7 A666]
QR201.A72A73   1986   616.9′2   86-4216
ISBN 0-8493-4504-9 (set)
ISBN 0-8493-4505-7 (vol. 1)
ISBN 0-8493-4506-5 (vol. 2)

International Standard Book Number 0-8493-4504-9 (Set)
International Standard Book Number 0-8493-4505-7 (Volume I)
International Standard Book Number 0-8493-4506-5 (Volume II)

Library of Congress Card Number 86-4216
Printed in the United States

PREFACE

In the past decade, biology has exploded into new areas of technology with promises of benefits to immunology, genetics, medicine, and agriculture. Replicating in vitro systems of vector cells have been developed in the last 20 years that could aid substantially in the application of the new tools to solutions of problems in these disciplines. These systems of vector cells provide a useful means with which to study the biological interactions of arthropod-borne animal viruses (arboviruses) and cells of their intermediate hosts, and in addition offer a sensitive device for the isolation, identification, and characterization of arboviruses. These two volumes are intended to provide a comprehensive reference dealing with the status of arthropod cell cultures as systems for arbovirus growth. Beginning with an elaboration of the historical events that led to the development and virological application of arthropod tissue and cell cultures, chapters are presented that encompass methodology for cultivation of vector tissues and cells, characterization and identification of vector cell lines, and isolation, replication, and diagnosis of arboviruses in arthropod cell systems. Recent research findings concerning specific applications of vector cell systems to problems in infectious disease, both in the laboratory and in field situations (e.g., yellow fever and dengue viruses), are presented. Following this, the promise and limitations of arthropod cell systems as substrates for the production of diagnostic reagents and vaccines are discussed. Finally, the replicative dynamics of arboviruses in arthropod cell systems are reviewed. Included are discussions of persistence, fusion, host-range mutants, and the genetic influence of arthropod cells on virus growth.

The editor wishes to thank the numerous contributors to this volume, without whose cooperation the book would not have been possible. Special thanks are due to Charlotte A. Miller, whose careful attention to processing of the manuscripts is greatly appreciated and to Dr. Jane E. Homan, Dr. David Stiller, and Dr. Douglas M. Watts who kindly read parts of this book and offered their helpful comments.

Conrad E. Yunker

FOREWORD

Arthropod-borne viruses (arboviruses) are among the most versatile agents known. They replicate at ambient temperatures in their arthropod vectors and at much higher temperatures in their vertebrate hosts. They subsist meekly for long periods, sometimes years, in the invertebrate host. By contrast they may cause explosive disease in the vertebrate host, sometimes resulting in the death of the animal. The same effects hold in cell culture. The effect in arthropod cells is usually nonpathogenic, while that in vertebrate cells is often destructive. This moderation in the arthropod cell has long been a subject for intellectual discussion, but only recently a subject for serious scientific exploration resulting in surprisingly enlightening findings. These new advances are now encapsulated in this timely treatise.

There are well over 400 arboviruses. Among these are some of the earliest recognized viruses. Yellow fever in the family Flaviviridae was shown at the turn of the century by Walter Reed and his colleagues to be mosquito-borne. The early studies in *Aedes aegypti* established arthropods as major vectors of viruses. Also at the turn of the century scientists in South Africa described bluetongue of sheep and later African horsesickness caused by orbiviruses in the family Reoviridae. These arboviruses are major pathogens of domestic animals and are transmitted by *Culicoides* midges. In 1910, R. E. Montgomery isolated African swine fever virus, the only DNA-containing arbovirus. This tick-borne virus has spread from Africa to the Americas and now poses a serious economic threat to the swine industry. By 1925, the Indiana serotype of vesicular stomatitis virus of the family Rhabdoviridae was recognized as a disease agent of livestock. Still more mosquito-borne viruses were revealed in the 1930s: Rift Valley fever in the family Bunyaviridae, and Venezuelan, Eastern, and Western equine encephalomyelitis viruses of the family Togaviridae. By 1950, 35 arboviruses were known. These encompassed human and domestic animal pathogens of each of the major genera in the five largest families of arboviruses.

The rapid increase after 1950 in knowledge of arboviruses and their vertebrate hosts was accompanied by widespread experimentation in vertebrate cell cultures. The comparable studies with invertebrate cells were not to come for almost 2 more decades.

This lag is partly explained by the lack of easily propagated arthropod cells, but is also explained by the bandwagon effect in a major clan of virologists. These were molecular virologists making spectacular discoveries at a basic level using vertebrate cells with arbovirus models such as vesicular stomatitis, Semliki Forest, and Sindbis viruses. So much was happening so fast that few investigators paused to reflect that viruses such as vesicular stomatitis are basically arthropod viruses, not vertebrate animal viruses. An arbovirus infects its vertebrate host only briefly, just long enough to produce viremia so that the vector can be infected. The vector, in contrast, is infected for life, and indeed sometimes (as in the case of the sandfly with Indiana strain of vesicular stomatitis) has the potential to maintain the virus indefinitely by vertical transmission. It is ironic that much of the basic molecular virology knowledge was accumulated in the vertebrate cell which is only visited briefly in the life cycle of arboviruses in nature.

In the 1960s a handful of pioneers such as J. Peleg, T. D. C. Grace, J. Rehacek, K. R. P. Singh, K. Maramorosch, S. M. Buckley, M. G. R. Varma, M. Pudney, I. Schneider, and C. E. Yunker (with their collaborators) developed first mosquito, then tick cell lines, and just as importantly, they developed simplified media and propagation techniques. Later, A. Igarashi cloned the *Aedes albopictus* line of Singh, not only providing a tool with remarkably increased ability to produce progency virus, but also confirming the basic tenet that not all cells in a population are born equal. This single technology applied to dengue and other arboviruses has enabled the large-scale reliable replication so necessary for molecular studies.

This book tells for the first time in one volume, the complete story of both the theoretical and practical application of this recently developed methodology. We have answers (or at least the means to obtain the answers) for questions such as "How is an arbovirus able to replicate in cells so divergent as vertebrate and arthropod?", "What are the factors limiting the production of virus in arthropod cells?", and "What determines the specificity of a given arbovirus for a given arthropod cell?".

The arthropod cells have been a boon to those wishing to isolate arboviruses in nature. Since the cells are maintained at temperatures approaching ambient in tropical climates, the cells can be carried in the pocket, inoculated in the field, and sent to the laboratory without special handling. It is no longer necessary to maintain a cold-chain for shipping field specimens suspected of containing an arbovirus. In most cases the arthropod cell, once infected, remains persistently infected, maintaining the culture of virus for weeks if necessary.

An unexpected dividend is now resulting from the use of mosquito cells for primary isolation of viruses from mosquitoes. A large number of viruses has recently been isolated which previously went unrecognized on inoculation of vertebrate cell cultures or animals. Most of these agents are not yet identified. They may well be viruses indigenous to mosquitoes and of no public health consequence, but the parallel is drawn to the isolation in the 1950s and 1960s of large numbers of new viruses from arthropods in baby mice. Some of these such as Crimean-Congo hemorrhagic fever and Oropouche viruses were initially thought to be of minor public health importance, but were subsequently shown to be major human pathogens. The human exposure to arthropods and to arthropod cells (some presumably infected with these viruses) can be a daily event, especially in the tropics, and the potential of the new mosquito cell isolates should not be disregarded.

The development and use of tick cells in culture have not come as easily or as quickly as the use of mosquito cells. As the technology progresses, however, this field will inevitably blossom and offer new vistas for study. The young investigator should take special note of tick cell culture as an area to be exploited.

Dr. Yunker is to be congratulated on bringing together the information on arthropod cells and arboviruses. There is yet much to be done and learned in this fascinating system. These two volumes will serve as a catalyst at this time of rapidly advancing technology.

Robert E. Shope, M. D.
Director
Yale Arbovirus Research Unit
Department of Epidemiology and
 Public Health
School of Medicine
Yale University
New Haven, Connecticut

CONTRIBUTORS

Danielle Blondel
Scientific Researcher
Laboratoire de Génétique des Virus
Centre National de la Recherche
  Scientifique
Gif-Sur-Yvette, France

Francoise Bras, D.Sc.
Assistant Chief
Laboratoire de Génétique des Virus
Centre National de la Recherche
  Scientifique
Gif-Sur-Yvette, France

Susan E. Brown, Ph.D.
Associate Research Scientist
Department of Epidemiology and
  Public Health
School of Medicine
Yale University
New Haven, Connecticut

D. Contamine, Dr.Sc.
Scientific Researcher
Laboratoire de Génétique des Virus
Centre National de la Recherche
  Scientifique
Gif-Sur-Yvette, France

Sybille Dezélée, D.Sc.
Scientific Researcher
Laboratoire de Génétique des Virus
Centre National de la Recherche
  Scientifique
Gif-Sur-Yvette, France

Kenneth H. Eckels, Ph.D.
Assistant Chief
Department of Biologics Research
Walter Reed Army Institute of
  Research
Washington, DC

Peter-Joachim Enzmann, Dr.rer.nat.
Wissenschaftlicher Oberrat
Federal Research Centre for Virus
  Diseases of Animals
Tübingen, West Germany

Duane J. Gubler, Sc.D.
Chief
Dengue Branch and
Director
San Juan Laboratories
Center for Infectious Diseases
Division of Vector-Borne Viral Diseases
Centers for Disease Control
San Juan, Puerto Rico

Akira Igarashi, M.D.
Professor and Chairman
Department of Virology
Institute for Tropical Medicine
Nagasaki University
Nagasaki, Japan

Martine Jozan, M.D., D.R.P.H.
Research Virologist
Adjunct Lecturer
School of Public Health
University of California at Los Angeles
Los Angeles, California

Christoph Kempf, Ph.D.
Institute of Hygiene and Microbiology
University of Bern
Bern, Switzerland

Dennis L. Knudson, D. Phil.
Associate Professor of Epidemiology
Department of Epidemiology and
  Public Health
Yale University
New Haven, Connecticut

Hans Koblet, M.D.
Professor
Institute of Hygiene and Microbiology
University of Bern
Bern, Switzerland

Goro Kuno, Ph.D.
Research Microbiologist
San Juan Laboratories
Division of Vector-Borne Viral Diseases
Centers for Disease Control
San Juan, Puerto Rico

Christine Kurstak, M.D., Ph.D.
Head
Laboratory of Virology
Hospital Hôtel-Dieu de Montréal
University of Montreal
Montreal, Québec, Canada

Edouard Kurstak, Dr. d'Etat, Ph.D.
Professor and Director
Department of Microbiology and
  Immunology
Faculty of Medicine
University of Montreal
Montreal, Québec, Canada

Colin J. Leake, Ph.D.
Lecturer
Department of Entomology
London School of Hygiene and
  Tropical Medicine
London, England

Helmut Mahnel, Dr. med. vet.
Professor
Faculty of Veterinary Medicine
University of Munich
Munich, West Germany

Eberhard Munz, Dr. med. vet.
Professor
Faculty of Veterinary Medicine
University of Munich
Munich, West Germany

Adames Omar, Ph.D.
Institute of Hygiene and Microbiology
University of Bern
Bern, Switzerland

Anne-Marie Petitjean
Ingeneer
Laboratoire de Génétique des Virus
Centre National de la Recherche
  Scientifique
Gif-Sur-Yvette, France

Mary Pudney, Ph.D.
Research Scientist
Department of Biochemical
  Microbiology
Wellcome Research Laboratories
Beckenham, Kent, England

Josef Reháček, Ph.D., D.Sc.
Research Scientist
Department of Rickettsiae
Institute of Virology
Slovak Academy of Sciences
Bratislava, Czechoslovakia

Max Reimann, Techn. Assist.
Faculty of Veterinary Medicine
University of Munich
Munich, West Germany

Imogene Schneider, Ph.D.
Department of Entomology
Walter Reed Army Institute of
  Research
Washington, DC

Robert McNair Scott, M.D., F.A.A.P.*
Head
Department of Virology
NAMRU-3
FPO New York, New York

Victor Stollar, M.D.
Professor
Department of Molecular Genetics and
  Microbiology
Robert Wood Johnson Medical School
University of Medicine & Dentistry of
  New Jersey
Piscataway, New Jersey

Danielle Teninges, D.Sc.
Scientific Researcher
Laboratoire de Génétique des Virus
Centre National de la Recherche
  Scientifique
Gif-Sur-Yvette, France

* Present affiliation: Assistant to Dirertor, Division of Communicable Disease and Immunology, Walter
  Reed Army Institute of Research, Washington, D.C.

Peter Tijssen, Ph.D.
Research Associate
Institut Armand-Frappier
Laval des Rapides,
Québec, Canada

M. G. R. Varma, Ph.D., D.Sc.
Professor of Medical Entomology
Department of Entomology
London School of Hygiene and
 Tropical Medicine
London, England

Francoise Wyers, D.Sc.
Scientific Researcher
Laboratoire de Génétique des Virus
Centre National de la Recherche
 Scientifique
Gif-Sur-Yvette, France

To Samira

TABLE OF CONTENTS

Volume I

INTRODUCTION
Chapter 1
Of Arboviruses, Arthropods, and Arthropod Cell Cultures:
History and Expectations 3
Martine Jozan

**CULTIVATION AND CHARACTERIZATION OF ARTHROPOD CELLS AS
SUBSTRATES FOR VIRUS GROWTH**
Chapter 2
Preparation and Maintenance of Arthropod Cell Cultures: Diptera, with
Emphasis on Mosquitoes 25
Imogene Schneider

Chapter 3
Preparation and Maintenance of Arthropod Cell Cultures: Acari, with
Emphasis on Ticks 35
Conrad E. Yunker

Chapter 4
Characterization and Identification of Arthropod Cell Lines 53
S. E. Brown and D. L. Knudson

Chapter 5
In Vitro Immunoenzymatic Detection and Screening of Arthropod-Borne
Togavirus Antigens and Antibodies 67
Edouard Kurstak, Peter Tijssen, and Christine Kurstak

**ARTHROPOD CELL CULTURES AS SYSTEMS FOR THE ISOLATION
AND ASSAY OF ARBOVIRUSES**
Chapter 6
Application of *Aedes pseudoscutellaris* (AP-61) Cells to Arbovirus Isolation and
Identification 79
C. J. Leake and M. G. R. Varma

Chapter 7
Tick Cell Lines for the Isolation and Assay of Arboviruses 87
Mary Pudney

Chapter 8
Application of *Aedes albopictus* Clone C6/36 Cells to the Isolation of
Mosquito-Borne Togaviruses in Japan, Indonesia, and Thailand 103
Akira Igarashi

Chapter 9
Arthropod Cell Cultures in Studies of Tick-Borne Togaviruses and Orbiviruses
in Central Europe 115
Josef Reháček

Chapter 10
Nairobi Sheep Disease Virus and *Reovirus*-Like Particles in the Tick Cell
Line TTC-243 from *Rhipicephalus appendiculatus:* Experiences
with the Handling of the Tick Cells, Immunoperoxidase, and
Ultrahistological Studies 133
Eberhard Munz, Max Reimann, and Helmut Mahnel

INDEX 149

Volume II

APPLICATIONS OF ARTHROPOD CELL CULTURES TO DENGUE
VIRUS IDENTIFICATION AND VACCINE PRODUCTION
Chapter 11
Application of Serotype-Specific Monoclonal Antibodies for Identification
of Dengue Viruses 3
D. J. Gubler

Chapter 12
Arthropod Cell Lines as Substrates for the Production of Dengue Virus
Vaccines...15
Kenneth H. Eckels and Robert McNair Scott

REPLICATIVE DYNAMICS OF ARBOVIRUSES IN ARTHROPOD
CELL SYSTEMS
Chapter 13
Comparative Growth of Arboviruses in Cell Lines Derived from *Aedes* and
Anopheles Mosquitoes and from the Tick *Boophilus microplus* 25
C. J. Leake

Chapter 14
Arbovirus Replication in Cell Cultures of Nonhematophagous Mosquitoes............43
Goro Kuno

Chapter 15
A Comparative View of the Replication of Togaviruses in Vertebrate and
Mosquito Cells 53
P.-J. Enzmann

Chapter 16
Persistent Infections of Mosquito Cells by Togaviruses 67
P.-J. Enzmann

Chapter 17
Fusion of Alphavirus-Infected Mosquito Cells 77
Hans Koblet, Adames Omar, and Christoph Kempf

Chapter 18
Arbovirus-Vector Cell Interactions and Host-Range Viral Mutants 91
Victor Stollar

Chapter 19
Rhabdoviruses in *Drosophila* Cells 111
S. Dezélée, F. Wyers, D. Blondel, A. M. Petitjean, D. Contamine, F. Bras,
and D. Teninges

INDEX 135

Applications of Arthropod Cell Cultures to Dengue Virus Identification and Vaccine Production

Chapter 11

APPLICATION OF SEROTYPE-SPECIFIC MONOCLONAL ANTIBODIES FOR IDENTIFICATION OF DENGUE VIRUSES

D. J. Gubler

TABLE OF CONTENTS

I.   Introduction 4

II.   Current Methods for Dengue Virus Identification 4
      A.   Plaque Reduction Neutralization Test 4
      B.   Complement-Fixation Test 5
      C.   Serotype-Specific Monoclonal Antibodies 5

III. Use of Monoclonal Antibodies with Different Virus Isolation Systems 6
      A.   Arthropod Cell Cultures 6
            1.   C6/36, *Aedes albopictus* 6
            2.   AP-61, *Aedes pseudoscutellaris* 7
            3.   TRA-284, *Toxorhynchites amboinensis* 7
            4.   RML-12, *Aedes aegypti* 7
      B.   Mammalian Cell Cultures 7
            1.   LLC-MK$_2$ 7
            2.   Vero 8
      C.   Mosquito Tissue 8

IV.   Specificity of Monoclonal Antibodies 9

V.   Sensitivity of Monoclonal Antibodies 13

VI.   Summary and Conclusion 13

References 13

I. INTRODUCTION

Dengue fever is the most important arbovirus disease of humans both in terms of morbidity and mortality,[1] with an estimated 1.5 billion people at risk of infection.[2] The marked increase in epidemic dengue activity in most parts of the tropics in recent years can be attributed primarily to increased air travel, which provides the ideal mechanism for the transport of dengue viruses between population centers of the world.

There are four serotypes of dengue viruses, dengue-1, -2, -3, and -4, and although they are antigenically related, there is no cross-protective immunity between the four viruses. A person can therefore conceivably have up to four separate dengue infections. Because the viruses are antigenically related, considerable cross-reactions occur in most serologic tests, especially if the infection is secondary or tertiary, thus making identification of the infecting dengue viruses difficult. This, plus the fact that dengue viruses have historically been among the most difficult of the arboviruses to isolate, has created many problems for the dengue diagnostic laboratory. As a result, only in recent years have dengue viruses been routinely isolated and identified by virus laboratories.

The first reliable method of dengue virus identification was the plaque reduction neutralization test (PRNT).[3] This test is highly specific and is still used routinely in some laboratories. With the introduction of mosquito inoculation,[4] which for the first time provided a highly sensitive method of dengue virus isolation, the complement-fixation (CF) test using antigen prepared in mosquitoes became widely used for identification.[5] Most dengue laboratories, until recently, used either PRNT or CF, but both tests require certain expertise that is usually not available. As a result, only a few laboratories in dengue endemic areas have had the capability to isolate and identify dengue viruses. Most of these laboratories placed primary emphasis on dengue serology. The recent development of serotype-specific monoclonal antibodies[6,7] and the use of highly sensitive mosquito cell cultures for virus isolation[8-10] have now made possible the routine isolation and identification of dengue viruses by most laboratories in endemic areas (see Chapters 6, 8, and 14).

II. CURRENT METHODS FOR DENGUE VIRUS IDENTIFICATION

A. Plaque Reduction Neutralization Test

The PRNT provides a highly reliable, serotype-specific method of identifying dengue viruses.[3,11] For this test viruses can be isolated in any system, but subsequent passage into susceptible mammalian cell cultures such as LLC-MK$_2$ (rhesus monkey kidney) cells is necessary. The test usually works best if viruses are passaged once in LLC-MK$_2$ cells under liquid medium to amplify the virus titer.[11] The virus is tested by PRNT using twofold dilutions of hyperimmune mouse ascitic fluid (HIMAF) prepared against each of the 4 dengue serotypes. Serum-virus mixtures are inoculated into LLC-MK$_2$ cell monolayers and after 6 days of incubation and 2 agar overlays, the last containing 1:7500 neutral red stain, plaques are counted on day 7. Either 50 or 90% plaque reduction end points can be counted.

Although the PRNT is a highly specific method of dengue virus identification, it has certain disadvantages. Most important, LLC-MK$_2$ cells have low susceptibility to dengue virus infection and unadapted viruses frequently do not replicate to titers sufficiently high on the first passage to be used in PRNT. New virus isolates, therefore, usually have to be passaged several times to allow adaptation to LLC-MK$_2$ cells, thus providing sufficient virus titers for identification. Another related disadvantage is that most laboratories that use the PRNT system for identification also use the LLC-MK$_2$ cells for dengue virus isolation. The lack of sensitivity of these cells to dengue virus infection means that only those viruses that are more adaptable to growth in LLC-MK$_2$ cells are isolated and identified.

B. Complement-Fixation Test

The CF test has been used to identify dengue virus isolates for many years, but it became widely used only after it was adapted for use with an antigen prepared from infected mosquitoes.[5] In recent years with the development of more sensitive mosquito cell lines for dengue virus isolation, the antigen prepared from these cell cultures has also been successfully used for CF identification of dengue viruses.

The standard test procedure developed at the Centers for Disease Control (CDC) is generally used with a few minor exceptions. The antigen which is prepared by triturating infected mosquitoes in borate saline, pH 9.0, is used undiluted without titration. The test uses 2.0 to 2.5 units of complement and HIMAF prepared against each of the four dengue serotypes as antisera. Since cross-reactions between the four dengue serotypes and other flaviviruses do occur, the results are often equivocal in some laboratories. Once the patterns of cross-reaction are determined for a particular lot of HIMAF, however, identification of dengue viruses is highly reliable. The major disadvantages of this test are its complexity and cross-reactions, which are sometimes difficult to interpret. The major advantage is that it does not require sophisticated equipment and is economical and therefore ideal for many laboratories with limited budgets.

C. Serotype-Specific Monoclonal Antibodies

Dengue virology has lagged behind other fields because of these difficulties associated with virus isolation and identification. The development by the Walter Reed Army Institute of Research (WRAIR)[6] of hybridomas that produce serotype-specific monoclonal antibodies to each of the four dengue serotypes, however, has changed that. The hybridomas were given to the Division of Vector-Borne Viral Diseases, Center for Infectious Diseases, CDC, for production and distribution, and in the past 2 years these antibodies have been distributed to most dengue laboratories in tropical regions. The use of monoclonal antibodies for identification and more sensitive mosquito cell cultures for isolation has provided a simple, economical, reliable, and rapid method for isolation and identification of dengue viruses that can be used in most virology laboratories.[7,12] The WRAIR group developed many hybridomas, but only a relative few were dengue virus serotype-specific (Table 1). All are available on request from CDC.

The monoclonal antibodies are usually used in an indirect fluorescent antibody test (IFAT),[7,12] but may also be used in an enzyme-linked immunosorbent assay.[13] Although a few problems still exist, the monoclonal antibodies can be used with viral antigen produced in a variety of cell cultures as well as in mosquito brain tissue.

To ensure that the monoclonal antibodies do not mix, Teflon-coated slides with 12 wells should be used. This allows two wells for each antibody in addition to the negative and positive controls. A line should be drawn between the wells of each antibody with a wax marker.

Monoclonal antibodies produced as ascites in mice usually have relatively high titers and have been used at dilutions as high as 1:300.[12] However they apparently lose potency at this high dilution (see below). The ascites should be used at a 1:40 dilution; the tissue culture fluid should be used at 1:10 dilution.

Table 1
SEROTYPE-SPECIFIC MONOCLONAL
ANTIBODIES AVAILABLE FOR DENGUE VIRUS
IDENTIFICATION[a]

Dengue serotype	Monoclonal antibody	IFAT activity
DEN-1	15F3	Poor reaction with some virus strains
	1F1	Good reaction with all viruses tested
DEN-2	3H5	Good reaction with all viruses tested
DEN-3	5D4	Good reaction with all but 1 virus tested
	8A1	Good reaction with all but 1 virus tested
DEN-4	1H10	Good reaction with all viruses tested

[a]   These antibodies are available on request from the Division of
Vector-Borne Viral Diseases, CDC, P.O. Box 2087, Fort Collins,
Colo. 80521.

III. USE OF MONOCLONAL ANTIBODIES WITH DIFFERENT VIRUS ISOLATION SYSTEMS

A. Arthropod Cell Cultures

In recent years several mosquito cell lines have been developed that are much more susceptible to dengue virus infection than other tissue culture cell lines. Those which have been used most widely by dengue laboratories include Igarashi's C6/36 clone of Singh's *Aedes albopictus* cell line,[8,12,14] Varma et al.'s AP-61 line from *A. pseudoscutellaris*,[9] and Kuno's TRA-284 line from *Toxorhynchites amboinensis*[10,12] (see also Volume I, Chapters 6 and 8 and Volume II, Chapter 14). Although inoculating live mosquitoes is still the most sensitive method of isolation, studies done in our laboratory showed that 88% of dengue viruses were isolated in the TRA-284-SF cell line.[18] This high isolation rate plus the ease and economics of processing large numbers of samples more than makes up for the lower sensitivity.

1. C6/36, Aedes albopictus

This cell line was one of the first to be widely used for dengue virus isolation.[8,12,14] A problem that has developed in recent years is that some lines of the clone have become less sensitive to dengue viruses, suggesting that it is not biologically homogeneous. While the original sensitivity has apparently been maintained in Igarashi's laboratory,[19] changes in the cells have been documented in ours and other laboratories.[10] In addition to decreased sensitivity, the C6/36 cells have not supported dengue virus replication as well as the AP-61 and TRA-284 lines.[10] Thus, even though a few cells became infected, some viruses do not replicate and spread to other cells in the culture. This has affected the ability to identify dengue viruses in C6/36 cells using the monoclonal antibody IFAT. For example, we processed human sera for virus isolation simultaneously in the C6/36, AP-61, and TRA-284 cell lines. The TRA-284 cells showed the highest susceptibility with 53 isolates (34%), compared with 51 isolates in AP-61 (33%), and only 43 in C6/36 cells (28%).[10] Because too few cells were infected, six viruses could not be identified in the initial cell culture isolation material and four of these were in the C6/36 cultures. These cultures had been incubated for 10 days after inoculation of the acute phase serum. Even so, less than 1% of the cells were infected, suggesting that the virus was not released from the infected cells. Whether this was due to viral or cell characteristics is not known, but the fact that these same viruses infected the TRA-284 cells and were identified without difficulty suggests that the cells were responsible. The infected cells in these cultures reacted well to the direct fluorescent antibody (DFAT) conjugate and to the HIMAF used as the positive control for the IFAT. However, there was not enough antigen to bind the more specific monoclonal antibodies.

The advantages of the C6/36 cell line were the ease with which the cells grew and could be handled and the cells were easily dispersed in monolayers on slides.

2. AP-61, *Aedes pseudoscutellaris*

The AP-61 cell line has been shown to be highly susceptible to infection with dengue viruses by several laboratories.[10,15] Infection is often determined by cytopathic effects (CPE) and virus isolates are usually identified by using the infected cells as antigen for the CF test. We have found the AP-61 cells more difficult to use for fluorescence microscopy. The cells rupture easily and clumping makes them difficult to disperse as monolayers on slides for IFAT or DFAT. The FA tests, therefore, are more difficult to read.

The AP-61 cells were more susceptible to dengue infection than the C6/36 cells.[10] Of 6 viruses not identified by monoclonal antibody IFAT in that experiment, 2 were in the AP-61 cells.

3. TRA-284, *Toxorhynchites amboinensis*

The TRA-284 parent line and the subline maintained on serum-free medium (TRA-284-SF) have proved to be the best cell lines overall for dengue virus isolation and identification.[10,16] The cells are highly susceptible to infection with these viruses and identification was accurate 100% of the time. The cells are easy to handle and very economical in that they do not require bovine serum in the growth or maintenance medium. The cells are larger than the C6/36 and screening for infected cells by DFAT is relatively easy.

4. RML-12, *Aedes aegypti*

The *Aedes aegypti* cell lines developed to date have not proved as susceptible to dengue virus infection as the cells from other species. Nevertheless they have had limited use, and virus titers of some prototype viruses in the RML-12 cell line were comparable to those observed in AP-61 cells.[17] Although this line has not been used for virus isolation, experimental studies have shown that the monoclonal IFA test works well for dengue virus identification in these cells.[18]

B. Mammalian Cell Cultures
1. LLC-MK₂

This cell line has had widespread use for isolation of dengue viruses. As noted above, however, LLC-MK₂ cells are less sensitive than the mosquito cell cultures to unadapted dengue viruses. Henchal et al.[7] reported that LLC-MK₂ cultures infected with dengue-2 virus (strain not reported) had 4+ fluorescence 24 hr p.i. and that overall fluorescence in these cells with both the 3H5 dengue 2 monoclonal antibody and a dengue-2 HIMAF was nearly as brilliant as it was in C6/36 cells infected at the same time. A similar study in our laboratory with unpassaged viruses, however, did not give similar results. Viruses of all four serotypes were inoculated undiluted into LLC-MK₂ and Vero (African green monkey kidney) cell cultures, incubated for 10 days at 36°C, and spotted on slides for fluorescent staining. All viruses had been previously isolated in TRA-284-SF cells and identified by monoclonal antibody IFAT.

Of 12 viruses inoculated, 10 were detectable in LLC-MK₂ cells by IFA test using

Table 2

COMPARATIVE IDENTIFICATION OF DENGUE VIRUSES GROWN IN TWO MAMMALIAN CELL LINES

		LLC-MK$_2$ cells		Vero cells	
Virus	Dengue serotype	Fluorescence	Identification	Fluorescence	Identification
1339	DEN-3	+	—	+	—
1351	DEN-1	—	NT	—	NT
1369	DEN-1	±	DEN-1	++	DEN-1
1412	DEN-1	—	NT	±	—
1413	DEN-1	±	—	—	NT
1416	DEN-3	++	DEN-3	++++	DEN-3
1420	DEN-4	+++	DEN-4	++++	DEN-4
1421	DEN-2	±	—	+++	DEN-2
1424	DEN-3	±	—	+	—
1428	DEN-1	++	DEN-1	++	DEN-1
1444	DEN-2	++++	DEN-2	++++	DEN-2
1448	DEN-4	++++	DEN-4	±	NT

Note: Indicates brilliance of fluorescence; ± indicates dull but specific fluorescence, while +++ and ++++ indicate very bright fluorescence; — indicates absence of specific fluorescence; and NT = no test.

HIMAF (Table 2). However, four of these cultures had only a few cells infected and were negative when tested with the monoclonal antibody IFAT. Thus, only six of the viruses were identified with the initial isolation material. In general, viruses in those cultures with 10% or more of the cells infected were easily identified by the monoclonal antibody IFAT.

2. Vero

This cell line has the same disadvantages as the LLC-MK$_2$ line in that unadapted dengue viruses do not grow well and large doses of virus are required to ensure infection of the cells. Thus, results with the 12 unpassaged viruses were similar. There were ten isolations, but only six viruses could be identified (Table 2). One virus (dengue-2) grew well in Vero but not in LLC-MK$_2$ cells, and another virus (dengue-4) grew well in LLC-MK$_2$ but not in Vero. As with the LLC-MK$_2$ cells, viruses in those cultures with 10% or more of the cells infected were easily identified by the monoclonal antibody IFAT. Thus, although there are no advantages to using mammalian cells with dengue viruses because of the lower sensitivity, there are no problems in using the monoclonal antibody IFAT to identify viruses that grow well in these cultures.

C. Mosquito Tissue

Because of the higher sensitivity, the mosquito inoculation technique is still the method of choice for dengue virus isolation attempts from important specimens such as those from dengue hemorrhagic fever patients or fatal cases. It may be convenient, therefore, to use the monoclonal antibody IFAT to identify dengue viruses in mosquito brain tissue.

Conflicting results have been reported. Beaty and his colleagues have had difficulty detecting specific dengue-1 fluorescence in mosquito brain tissue with this test.[20] However, we have correctly identified by monoclonal antibody IFAT several dengue-1 viruses isolated from human sera during an outbreak in Veracruz, Mexico, in 1982.[21] In addition, a blind trial with all 4 dengue serotypes has recently been carried out (Table 3). Each serotype was correctly identified and there were no false positives. It should

Table 3
IDENTIFICATION OF DENGUE VIRUSES IN MOSQUITO BRAIN TISSUE BY MONOCLONAL ANTIBODY IFAT

Monoclonal antibody	Antigen			
	DEN-1	DEN-2	DEN-3	DEN-4
Negative control	−	−	−	−
Positive control	+++	+++	+++	+++
DEN-1-15F3	+	−	−	−
DEN-2-3H5	−	+++	−	−
DEN-3-5D4	−	−	++	−
DEN-4-1H10	−	−	−	++

Note: Indicates brilliance of fluorescence; + indicates dull but specific fluorescence, while ++ and +++ indicate brighter fluorescence; and − indicates absence of specific fluorescence.

be noted, however, that there may be a considerable amount of nonspecific staining in mosquito brain tissue and an inexperienced person could misinterpret the results. It is therefore recommended that the monoclonal antibody IFAT be used with mosquito brain tissue only in an emergency, and results should always be confirmed by CF using antigen prepared from the mosquito bodies or by passage into mosquito cell cultures.

IV. SPECIFICITY OF MONOCLONAL ANTIBODIES

To date only two studies have reported use of monoclonal antibodies to identify large numbers of newly isolated dengue viruses. Henchal et al.[7] at WRAIR identified over 100 strains of dengue-1, -2, and -4 that had been isolated from mosquitoes and human sera in West Africa and Jamaica. They reported no discrepancies with those confirmed by PRNT.

In a more extensive effort we studied 561 dengue viruses of all four serotypes from all major regions of the tropics (Table 4).[12] All were identified with the monoclonal antibody IFAT. The identification of 89 representative viruses from the same group was checked by CF with antigens prepared in mosquitoes. There was 100% agreement (Table 4). Some of the viruses of all serotypes in these two studies were isolated over 35 years apart, but identification was still possible, suggesting a very low mutability of the type-specific determinants on dengue viruses.[7] Thus the monoclonal antibodies proved effective in identifying dengue viruses of all four serotypes from all major parts of the tropics.

Recent evidence, however, suggests that some dengue-1 and -3 viruses are not easily identified with the monoclonal antibodies. The first evidence of this was with a virus isolated from an American who had been traveling in Colombia. The acute-phase serum was processed for virus isolation in both mosquito cell culture and by mosquito inoculation.[18] A *Flavivirus* was isolated in both systems which reacted strongly with the DFAT conjugate. When processed for identification using the monoclonal antibody IFAT on TRA-284-SF cells, however, there was no reaction. The patient had high yellow fever CF and HI antibody titers and the possibility of yellow fever or another *Flavivirus* infection was considered (Table 5). Mosquitoes infected parenterally were used to prepare CF antigen which was tested against a number of *Flavivirus* antisera for virus identification. The results were compatible with dengue-1 infection (Table 5) which was subsequently confirmed by PRNT.

Table 4
GEOGRAPHIC ORIGINS AND SOURCE OF VIRUSES IDENTIFIED

Virus	Origin	Source of isolate	Passage history	Identification method[a]	
				Monoclonal antibody	Complement-fixation
DEN-1	Puerto Rico	Human serum	None	112	25
	Mexico	Human serum	None	7	7
	Venezuela	Human serum	None	1	1
	Indonesia	Human serum	None	3	3
	Sri Lanka	Human serum	None	1	1
Subtotal				124	37
DEN-2	Puerto Rico	Human serum	None	3	3
	Jamaica	Human serum	None	1	1
	Indonesia	Human serum	None	3	3
	Sri Lanka	Human serum	None	1	1
	Upper Volta	Human serum	None	2	2
Subtotal				10	10
DEN-3	Puerto Rico	Human serum	None	5	5
	Indonesia	Human serum	None	6	6
	Malaysia	Human serum	None	3	3
	Thailand	Human serum	SMB-5,Tox-2[b]	3	3
	Sri Lanka	Human serum	None	3	3
Subtotal				20	20
DEN-4	Puerto Rico	Human serum	None	406	21
	Indonesia	Human serum	None	1	1
Subtotal				407	22
Totals				561	89

[a] All viruses identified by CF were also identified by monoclonal antibodies. From Gubler, D. J. et al., Reference 12.

[b] SMB-5 = 5 passages in suckling mouse brain and Tox-2 = 2 passages in *Toxorynchites* mosquitoes.

Table 5
SEROLOGIC RESULTS WITH SERUM AND VIRUS ISOLATED FROM COLOMBIA PATIENT

	Serology				
	HI[a]		CF[b]		CF virus identification[c]
Antigen/antisera	Serum 1	Serum 2	Serum 1	Serum 2	
DEN-1	<10	320	<8	64	2048
DEN-2	<10	160	<8	8	256
DEN-3	<10	160	<8	128	512
DEN-4	<10	160	<8	256	64
Yellow fever (17D)	<10	1280	<8	512	<16
St. Louis encephalitis	<10	320	<8	512	32
Ilheus	NT	NT	NT	NT	<16
Eastern equine encephalitis	<10	<10	<8	<8	NT

[a] Hemagglutination-inhibition.
[b] Complement-fixation.
[c] Using antigen prepared from infected mosquitoes.

This was the first evidence that not all dengue viruses could be identified by using an IFAT with the monoclonal antibodies currently used by most laboratories. A repeat IFAT using a new vial of the 15F3 dengue-1 monoclonal antibody at a lower dilution (1:40) confirmed that the virus did react, but the brilliance of fluorescence was considerably lower than that with the control HIMAF and only a few of the many infected cells showed specific fluorescence.

The reasons for the lack of reactivity with the 15F3 monoclonal antibody, which had worked well with other dengue-1 viruses are not known. It is possible, however, that if there was a higher ratio of virion to nonvirion antigen in the cell cultures, the dengue-1 (15F3) and -3 (5D4) monoclonal antibodies would not work as well as the dengue-2 (3H5) and -4 (1H10) antibodies, because the 15F3 and 5D4 antibodies are directed against nonstructural viral antigen(s). This could explain why only a few of the many infected cells that reacted with the DFA conjugate reacted with the dengue-1 monoclonal antibodies. Second, repeated freezing and thawing may have caused the diluted monoclonal antibodies to lose their potency.[22] Subsequently several other dengue-1 viruses were isolated in our laboratory that did not react well with the 15F3 monoclonal antibody.

In an effort to clarify the relationship of the 15F3 dengue-1 monoclonal antibody to various dengue-1 virus strains, a second dengue-1 monoclonal antibody was obtained from Dr. M. K. Gentry of WRAIR. The latter, designated 1F1, was prepared against the structural antigen. Strains of dengue-1 from Colombia, Haiti, Puerto Rico, and Mexico, all of which could not be identified in the initial isolation culture with the 15F3 monoclonal antibody, were tested again by infecting TRA-284-SF cell cultures. The cultures were incubated for 10 days at 28°C and the cells were harvested and spotted on slides for testing with the dengue-1 monoclonal antibodies, both of which were prepared as ascitic fluid and used at a 1:40 dilution.

Approximately 90% of the cells were infected with the Colombian dengue-1 virus, as determined by both DFAT and the positive control HIMAF IFAT (Table 6). Fluorescence was brilliant in all cells. When the IFAT was done with the 15F3 monoclonal antibody, however, less than 10% of the cells reacted and in these the fluorescence was dull. By contrast, the fluorescence was as brilliant with the 1F1 monoclonal antibody as with the positive control HIMAF. Moreover, the 1F1 antibody reacted with all infected cells (90%).

Similar results were observed with the other dengue-1 strains tested (two from Haiti and one each from Puerto Rico and Mexico). Thus, only a small proportion of the cells infected with these viruses reacted with the 15F3 antibody and those cells that did react had a dull fluorescence compared with that of the positive IFAT control (Table 6). By contrast, all of the dengue-1 strains reacted just as strongly with the 1F1 antibody as they did with the positive control HIMAF. Furthermore the fluorescence in all infected cells was of comparable brilliance.

A similar problem of identification occurred with a strain of dengue-3 virus (1339) isolated during the 1977 epidemic in Puerto Rico. This virus was reisolated from the human serum by mosquito inoculation and identification was confirmed by CF. In tests with several monoclonal antibodies, however, very poor reactions were observed (Table 7). Both LLC-MK₂ and TRA-284-SF cells were infected and spotted on slides after 7 and 10 days of incubation, respectively. All of the TRA-284-SF cells were infected as demonstrated by the positive control HIMAF, compared with about 10% of the cells in LLC-MK₂ cultures. The *Flavivirus* complex antibody (4G2) reacted well with the virus in both cultures. Surprisingly none of the other monoclonal antibodies reacted well with the infected cells, including a dengue-complex antibody, D71B. Thus, even in the TRA-284-SF cultures only a few positive cells were detected in each well, even though all cells were infected.

Table 6
REACTION OF SELECTED STRAINS OF DENGUE-1 TO MONOCLONAL ANTIBODIES[a]

Monoclonal antibody	Geographic strains of Dengue-1			
	Colombia	Haiti	Puerto Rico	Mexico
Negative control	−[b]	−	−	−
Positive control	+++	+	+++	+
DEN-1 15F3	±	±	±	±
DEN-1 1F1	+++	+	+++	+
DEN-2 3H5	−	−	−	−
DEN-3 5D4	−	−	−	−
DEN-4 1H10	−	−	−	−

Note: − = Negative, ± = <10% of cells with specific fluorescence, + = 10 to 24% of cells with specific fluorescence, ++ = 25 to 49% of cells with specific fluorescence, +++ = 50 to 90% of cells with specific fluorescence, and ++++ = 100% of cells with specific fluorescence.

[a] Indirect fluorescent antibody test. All viruses were inoculated into tube cultures of TRA-284-SF simultaneously and incubated for 10 days at 280°C.

[b] Scoring system refers to the number of cells with specific fluorescence.

Table 7
REACTION OF SEVERAL *FLAVIVIRUS* MONOCLONAL ANTIBODIES TO A PUERTO RICO STRAIN OF DENGUE-3

Monoclonal antibody[a]	Cell culture	
	TRA-284-SF	LLC-MK₂
Negative control	−[b]	−
Positive control	++++	+
DEN complex—D71B	+	±
Flavi complex 4G2	+++	±
Yellow fever	−	−
DEN-3—D38A1	±	−
DEN-3—D69E1	±	−
DEN-3—D35C9	±	±

[a] Provided by Dr. M. K. Gentry of Walter Reed Army Institute of Research.

[b] See Table 6 for scoring system of fluorescence.

Because of the above problems, the following recommendations for use of the monoclonal antibodies are made. All tissue culture fluid monoclonal antibodies should be used at a 1:10 dilution. Although end point titers may vary among lots, use at 1:10 will ensure good results. End point titers on the vials should *not* be used. Second, lyophilized, monoclonal antibodies that have been eluted should be aliquoted and stored undiluted. The working dilution (1:10) should be made only at the time of use. All questionable isolates should be confirmed by either CF or PRNT.

V. SENSITIVITY OF MONOCLONAL ANTIBODIES

Experience gained to date has demonstrated that the major problem associated with monoclonal antibody identification of dengue viruses is poor viral replication in the cell culture isolation system. In cultures with only a few infected cells, identification is generally not possible; at least one passage is necessary to allow the virus infection to spread to more cells. In cultures with 10% or more of the cells infected, identification is usually very reliable. The use of biotinylated antimouse IgG and fluorescein-labeled Avidin D with the monoclonal antibody IFAT increases the sensitivity of the test and frequently allows identification of viruses with low cell infection without passage.[12] All slow-growing viruses should be passaged to facilitate identification and all questionable isolates should be confirmed by either CF or PRNT.

VI. SUMMARY AND CONCLUSION

Serotype-specific monoclonal antibodies have provided for the first time a rapid, simple, economical, and accurate method for the routine identification of dengue viruses that most laboratories will be able to use. The monoclonal antibody IFAT, used in conjunction with sensitive mosquito cell cultures, should make virologic surveillance for dengue viruses routine.

Nevertheless, caution should be exercised in view of the observations that not all dengue strains, in particular dengue-1 and -3, react well with the monoclonal antibodies commonly used. More experience is needed before we understand the full implications of the strain variation observed. As more laboratories begin using these monoclonal antibodies on a routine basis, the extent and type of virus variation will become more evident, especially if a battery of monoclonal antibodies directed at different epitopes is used.

REFERENCES

1. Rosen, L., Dengue — an overview, in *Viral Diseases in South-East Asia and the Western Pacific*, Mackenzie, J. S., Ed., Academic Press, Canberra, 1983, chap. 66.
2. Halstead, S. B., Dengue haemorrhagic fever — a public health problem and a field for research, *Bull. W.H.O.*, 58, 1, 1981.
3. Russell, P. K. and Nisalak, A., Dengue virus identification by the plaque reduction neutralization test, *J. Immunol.*, 99, 291, 1967.
4. Rosen, L. and Gubler, D. J., The use of mosquitoes to detect and propagate dengue viruses, *Am. J. Trop. Med. Hyg.*, 23, 1153, 1974.
5. Kuberski, T. T. and Rosen, L., Identification of dengue viruses using complement fixation antigen produced in mosquitoes, *Am. J. Trop. Med. Hyg.*, 26, 538, 1977.
6. Gentry, M. K., Henchal, E. A., McCown, J. M., Brandt, W. E., and Dalrumple, J. M., Identification of distinct determinants on dengue-2 virus using monoclonal antibodies, *Am. J. Trop. Med. Hyg.*, 31, 548, 1982.
7. Henchal, E. A., McCown, J. M., Seguin, M. C., Gentry, M. K., and Brandt, W. E., Rapid identification of dengue virus isolates by using monoclonal antibodies in an indirect immunofluorescence assay, *Am. J. Trop. Med. Hyg.*, 32, 164, 1983.
8. Igarashi, A., Isolation of Singh's *Aedes albopictus* cell clone sensitive to dengue and chikungunya viruses, *J. Gen. Virol.*, 40, 530, 1978.
9. Varma, M. G. R., Pudney, M., and Leake, C. J., Cell lines from larvae of *Aedes (Stegomyia) malayensis* Corless and *Aedes (S.) pseudoscutellaris* (Theobald) and their infection with some arboviruses, *Trans. R. Soc. Trop. Med. Hyg.*, 68, 374, 1974.
10. Kuno, G., Gubler, D. J., Velez, M., and Oliver, A., Comparative sensitivity of three mosquito cell lines for isolation of dengue viruses, *Bull. W.H.O.*, 63, 279, 1985.

11. Bancroft, W. H., McCown, J. M., Mas Lago, P., Brandt, W. E., and Russell, P. K., Identification of dengue viruses from the Caribbean by plaque-reduction neutralization test, in *Dengue in the Caribbean, 1977,* Publ. No. 375, Pan American Health Organization Scientific, Washington, D.C., 1979, 173.
12. Gubler, D. J., Kuno, G., Sather, G. E., Vélez, M., and Oliver, A., Mosquito cell cultures and specific monoclonal antibodies in surveillance for dengue viruses, *Am. J. Trop. Med. Hyg.,* 33, 158, 1984.
13. Kuno, G., Gubler, D. J., and Santiago DeWeil, N., Antigen capture ELISA for the identification of dengue viruses, *J. Virol. Methods,* 12, 93, 1985.
14. Tesh, R. B., A method for the isolation and identification of dengue viruses using mosquito cell cultures, *Am. J. Trop. Med. Hyg.,* 28, 1053, 1979.
15. Race, M. W., Williams, M. C., and Agostini, C. F. M., Isolation of dengue virus in the *Aedes pseudoscutellaris* cell line (LSTM-AP-61), in *Dengue in the Caribbean, 1977,* Publ. No. 375, Pan American Health Organization Scientific, Washington, D.C., 1979, 165.
16. Leake, C. J., Nisalak, A., and Burke, D. S., Comparative isolation of dengue viruses from DHF patients by mosquito inoculation and on three mosquito cell lines, in *Proceedings on International Conference on Dengue and Dengue Hemorrhagic Fever,* Pang, T. and Rathmanathan, R., Eds., Kuala Lumpur, Malaysia, 1983, 437.
17. Kuno, G., Cultivation of mosquito cell lines in serum-free media and their effects on dengue virus replication, *In Vitro,* 19, 707, 1983.
18. Gubler, D. J. and Kuno, G., Unpublished data.
19. Igarashi, A., Personal communication.
20. Beaty, B. J. et al., Personal communication.
21. Gubler, D. J. and Vélez, M., Unpublished data.
22. Henchal, E. A., Personal communication.

Chapter 12

ARTHROPOD CELL LINES AS SUBSTRATES FOR THE PRODUCTION OF DENGUE VIRUS VACCINES

K. H. Eckels and R. McN. Scott

TABLE OF CONTENTS

I. Introduction 16

II. Replication of Dengue Viruses in Mosquito Cell Lines 16

III. Passage of Dengue Viruses in the C6/36 Cell Line 16

IV. Safety Tests Performed on Mosquito Cell Lines 19
    A. Tests for Adventitious Agents 19
    B. Tumorigenicity 19
    C. Karyologic and Isoenzyme Analyses 19
    D. Skin Tests in Monkeys and Man 19

V. Conclusion 20

Acknowledgments 20

References 21

I. INTRODUCTION

Mosquito cell lines first established by Singh[1] from *Aedes albopictus* and *A. aegypti* larvae were later shown to be useful for the replication of dengue (DEN) and other arboviruses.[2-6] Some advantages of these lines include adaptability to a variety of culture media, rapid growth, ability to be maintained at a wide range of temperatures, growth in suspension culture, and sensitivity for virus isolation. Igarashi, starting with the SAAR line of Singh's *A. albopictus* cells, isolated 20 clones in the presence of chikungunya (CHIK) virus antiserum.[7] During passage the clones also received treatment with DEN types 1, 2, 3, and 4 antisera. These antisera were used as a precautionary measure by Igarashi after finding that another line of Singh's *A. albopictus* cells (SAAK) was contaminated with a variant of CHIK virus. The clone designated as C6/36 was found to be more sensitive than others for replication of DEN and CHIK viruses and could be used to prepare high-titered virus stocks. The C6/36 cell clone was later used by Igarashi to establish persistent infections with all four serotypes of DEN virus.[8] Harvests from the persistently infected cultures contained a major population of temperature sensitive virus which was also found to be less pathogenic than standard DEN viruses for suckling mice.

The capacity of the C6/36 *A. albopictus* clone to generate high titers of DEN viruses while selecting for temperature sensitive variants on prolonged culture led to the investigation of this cell line for possible use as a vaccine substrate. Other cell lines that were studied were those derived from the nonbiting mosquito *Toxorhynchites amboinensis*.

II. REPLICATION OF DENGUE VIRUSES IN MOSQUITO CELL LINES

It has been shown by Singh[4] and others that DEN and other arboviruses from various sources are able to replicate in the *Aedes albopictus* cell line and this is true as well for the C6/36 clone. Adaptation and blind passage are usually not required with this cell line for DEN viruses originating in mouse brain, mammalian cell cultures, mammalian blood, or mosquitoes. In most cases recovery and replication are more successful than using mammalian cell lines.[9]

The replication of vaccine candidate strains DEN-2 PR-159 and DEN-3 CH53489 in both C6/36 and TA-42 cells are shown in Figure 1. Both of these strains originated as human serum isolates which were passaged 4 to 6 times in primary monkey kidney cell cultures. Maximum titers for these viruses were obtained 6 to 8 days in both mosquito cell lines. Superior growth in the C6/36 line was common for these DEN strains and also DEN types 1 and 4 with virus titers reaching 10^6 to 10^7 PFU/mℓ. These data are compatible with those of Igarashi[7] who also demonstrated high yields of DEN viruses in the C6/36 clone. These yields are commonly 10 to 100 times higher than those obtained from mammalian cell lines inoculated with the same DEN viruses.[10]

III. PASSAGE OF DENGUE VIRUSES IN THE C6/36 CELL LINE

Passage in cell cultures has been employed on numerous occasions to attenuate viruses so that they may be used for immunization. Currently available live-attenuated vaccines for yellow fever[13] and rubella[14] were produced in this manner. The basis for the attenuation is not known, but Holland[15] suggests that certain cell systems promote rapid genome evolution (especially the RNA viruses) that results in multiple genome changes in these viruses. These genome changes may be reflected in the attenuation characteristics of these viruses when they are presented to the human or animal recipient.

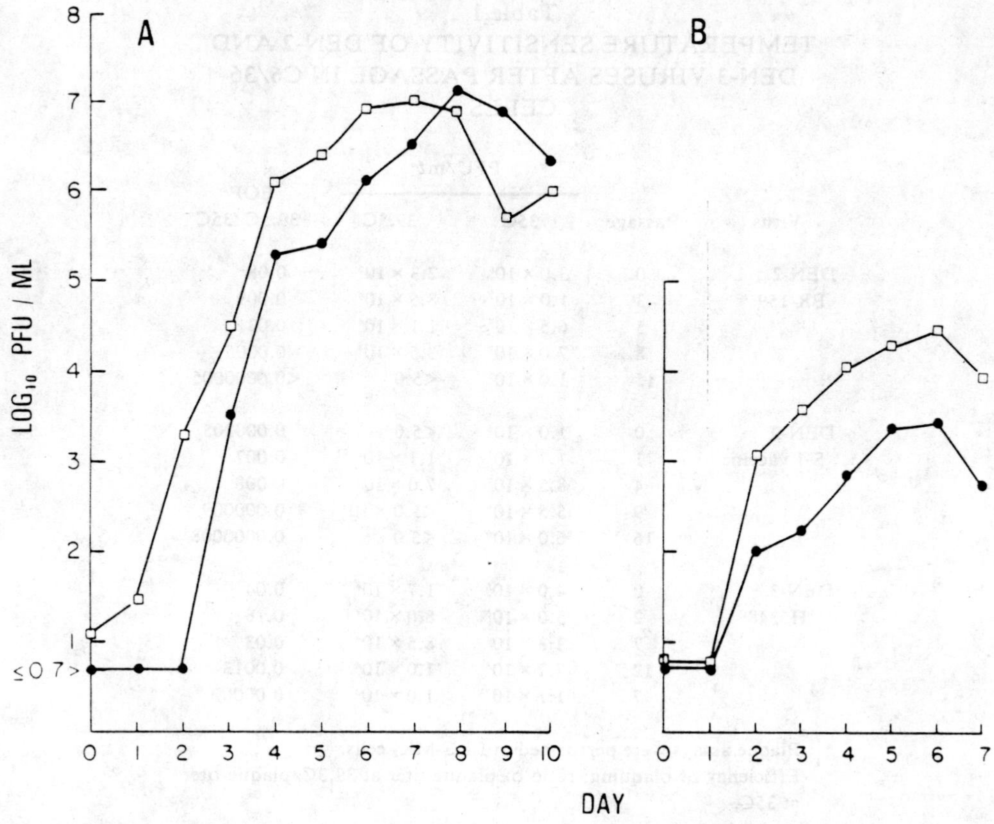

FIGURE 1.   Replication of DEN-2 PR-159 (•) and DEN-3 CH53489 (□) viruses in A. C6/36 cells and B. TA-42 cells.

Temperature sensitivity is a laboratory marker associated with attenuation for many viruses. Selection of DEN virus temperature sensitive (ts) mutants by establishment of persistently infected C6/36 cells has been demonstrated by Igarashi.[8] By serial passage of DEN-2 and DEN-3 vaccine candidate strains in C6/36 cells we were able to also select for ts virus populations. Table 1 lists the results of weekly passage of these uncloned virus strains. For both uncloned vaccine candidate viruses ts populations became evident between passages 8 and 12 with a steady increase in this population on subsequent passages. For the S-1 vaccine which is a clone of DEN-2 PR-159, a somewhat different pattern of selection occurred (Table 1). A very rapid destabilization of the clone was evident after only one passage in C6/36 cells with the emergence of a temperature resistant population of virus that could be plaqued at 39.3°C. Not until the ninth passage was there selection for ts viral populations. The small plaque ts virus found in C6/36 passage 9 was similar to the S-1 clone that was used to initiate the passages.

The DEN-2 PR-159/S-1 vaccine produced in fetal rhesus monkey lung diploid fibroblasts has been shown to be attenuated for various animals and humans.[16-19] Animal data for the vaccine virus and its parent virus are shown in Table 2. After passage of the DEN-2 PR-159 virus in C6/36 cells, monkey virulence as measured by the number of days of viremia is still as high as the unpassaged virus. The immune response for both the DEN-2 PR-159 unpassaged and passaged viruses shows that the monkeys were infected and produced significant levels of neutralizing antibodies. Although the C6/36-15 passage of DEN-2 PR-159 is highly ts, both mouse neurovirulence and mon-

Table 1

TEMPERATURE SENSITIVITY OF DEN-2 AND DEN-3 VIRUSES AFTER PASSAGE IN C6/36 CELLS

		PFU/mℓ[a]		EOP[b]
Virus	Passage	35C	39.3C	39.3C/35C
DEN-2	0	3.0×10^6	2.3×10^5	0.08
PR-159	3	1.0×10^7	8.5×10^4	0.009
	5	6.5×10^7	1.1×10^6	0.02
	8	7.0×10^7	3.5×10^4	0.0005
	15	1.0×10^7	<5.0	<0.0000005
DEN-2	0	1.0×10^6	<5.0	0.000005
S-1 vaccine	1	1.7×10^7	1.1×10^5	0.007
	4	8.5×10^7	7.0×10^5	0.008
	9	5.5×10^6	$<5.0 \times 10^1$	0.000009
	16	6.0×10^6	<5.0	0.0000008
DEN-3	0	4.0×10^5	1.7×10^4	0.04
CH53489	2	5.0×10^5	8.0×10^4	0.16
	7	3.3×10^6	8.5×10^4	0.03
	12	7.7×10^6	1.0×10^4	0.0013
	17	1.1×10^7	1.0×10^3	0.00009

[a] Plaque assays were performed in LLC-MK$_2$ cells.
[b] Efficiency of plaquing; ratio of plaque titer at 39.3C/plaque titer at 35C.

Table 2

MONKEY AND MOUSE VIRULENCE OF DEN-2 VIRUSES PASSAGED IN C6/36 CELLS

Virus	Passage	No. of monkeys	Inoculum (PFU)	Rhesus monkeys viremia (Mean no. of days)	Neutralizing antibody at 30 d (PRNT$_{50}$)	Suckling mouse LD$_{50}$/PFU
DEN-2	0	2	7.0×10^5	5.5	950[a]	0.13
PR-159	15	2	1.1×10^7	5.5	340	0.2
DEN-2	0	5	3.8×10^4	<1.0	60	0.002
S-1 vaccine	4	2	8.8×10^6	3.0	250	0.5
	9	2	2.2×10^6	0	130	ND
	13	2	6.0×10^6	0	<10	ND

[a] Reciprocal dilution of serum required to reduce a standard plaque dose by 50%.

key virulence are unchanged from the unpassaged virus. For other clones and passages of DEN-2 PR-159 virus, covariation of these markers has been observed.[16] Data for the DEN-2 PR-159 C6/36-15 virus indicate that separate areas of the viral genome might have been affected during passage to result in a segregation of these markers.

The S-1 vaccine virus passaged in C6/36 cells resulted in the emergence of a population of non ts virus soon after the initiation of passage which coincided with the selection of a virus population that was virulent for rhesus monkeys. From the data listed in Table 2, the S-1 vaccine virus after four passages in C6/36 cells could produce viremia in rhesus monkeys. Virus from the same passage could also kill suckling mice to a level approximately that of the DEN-2 PR-159 parent virus. Further passage in

C6/36 cells resulted in the selection of an attenuated virus population by the ninth passage, which was still infectious enough in the monkey to induce an immune response resulting in significant levels of neutralizing antibodies. Progressive attenuation with further passage of the virus in C6/36 cells was evident. By the 13th passage, two monkeys were free of viremia and did not seroconvert after inoculation with the S-1 virus, indicating that little or no infection had taken place. The measure of virulence for the suckling mouse, the LD_{50}/PFU ratio, was 0.0002, about 10-fold lower than the original unpassaged S-1 virus, also indicating that this virus was highly attenuated. Reversion to virulence with covariation of in vitro and in vivo markers for S-1 was evident in the first four C6/36 passages. Selection for ts beyond C6/36 passage four also resulted in selection for attenuation in rhesus monkeys and suckling mice.

IV. SAFETY TESTS PERFORMED ON MOSQUITO CELL LINES

The *Aedes albopictus* clone C6/36 cells were cultured as previously described.[7] Tests were performed in the cells to ensure that a product made in these cells was free of adventitious agents or substances that would elicit an untoward reaction in a human recipient. Many of the tests were those used for live-attenuated viral vaccines described in the Code of Federal Regulations (U.S.A.) pertaining to food and drugs.[11] *Toxorhynchites amboinensis* cell lines TA-9, TA-42, and TRA-284 were only tested for bacterial sterility and monkey skin test reactivity.

A. Tests for Adventitious Agents
Testing for microbial adventitious agents included the inoculation of C6/36 cell culture fluids and cells into bacterial and mycoplasma culture media, various insect and mammalian cells, and laboratory animals. The C6/36 line was also fused with other insect and mammalian cell cultures and observed for cytopathic effect. In no case was there any evidence of a demonstrable adventitious agent present in the C6/36 cells.

Additionally, electron microscopic studies were performed on thin sections of fixed C6/36 cells and observed for virus-like particles. None were seen in a total of 669 cells.

Reverse transcriptase possibly associated with retroviruses was assayed using concentrated C6/36 cell supernatant culture fluids. These assays were negative for reverse transcriptase activity.

B. Tumorigenicity
Newborn hamsters treated with antilymphocyte serum were inoculated with viable C6/36 cells and followed for the development of tumors. No tumors developed in these animals.

C. Karyologic and Isoenzyme Analyses
Stained chromosomes indicated a modal chromosome number of six in 82% of the cells studied. Isoenzyme patterns for C6/36 were the same as the *Aedes albopictus* line ATC-15 CCL126.

D. Skin Tests in Monkeys and Man
Sham vaccines were prepared from virus-free cell culture fluids taken from C6/36, TA-9, TA-42, and TRA-284 cell cultures. These cultures were washed free of FBS and the culture fluids removed after 7 days at 28°C and filtered through a 450 nm filter. Results of skin tests done in monkeys with all of the mosquito culture fluids and in humans using only C6/36 culture fluids indicated a high degree of hypersensitivity in these recipients resulting in immediate and delayed allergic responses.[10,12]

V. CONCLUSION

The *Aedes albopictus* clone C6/36 proved to be an excellent host cell line for the replication of vaccine strains of DEN-2 and DEN-3 viruses. Yields of these viruses in *Toxorhynchites amboinensis* mosquito lines were lower than in C6/36 and similar to those titers found when these viruses were grown in mammalian cell lines. For DEN-3 virus especially, the C6/36 cell line supported replication of this virus to levels which in many cases are not attainable in mammalian cell systems.

The DEN-2 PR-159/S-1 vaccine, prepared in fetal rhesus monkey diploid cells and titrated in yellow fever-immune humans was found to have an ID_{50} of $10^{3.3}$.[20] The ID_{50} dose in nonimmunes was not calculable because of insufficient seroconversions at the concentrations of the vaccine tested. In this case higher concentrations of the virus were required for infection and seroconversion. The 10- to 100-fold higher levels of virus found for vaccine strains grown in the C6/36 cell line would probably meet this requirement.

Passage of DEN-2 and DEN-3 vaccine strains in the C6/36 cell line generated ts populations of virus. In one case, passage of a cloned preparation, the DEN-2 S-1 vaccine, resulted in a rapid reversion to non ts virus followed by selection for small plaque, ts virus. The non ts passage contained a virus that was virulent for mice and monkeys, while later passages which were ts were also attenuated in animals. Continuous passage in C6/36 cells resulted in a progressively more attenuated virus population that was finally incapable of infecting monkeys. The DEN-2 PR-159 virus on the other hand was highly ts but not attenuated at a 15th passage level in C6/36 cells. Further passage in C6/36 cells may be required for animal attenuation. The lack of covariance of these two markers has been observed before on passage of DEN-4 virus in primary dog kidney cell cultures. In this case, monkey viremia as a measure of virulence was reduced at the 15th passage level while the emergence of ts virus populations did not occur until the 50th passage.[21]

A C6/36 sham vaccine prepared from virus-free cell culture fluids evoked immediate and delayed hypersensitivity responses in human volunteers.[12] Immediate hypersensitivity was also found in rhesus monkeys given the C6/36 sham vaccine and when similar preparations of *T. amboinensis* cell culture fluids were given.[10] The high incidence of reactivity to the mosquito cell preparations in both man and monkey indicates that a widespread sensitization must exist in these species.

The possible anaphylactic reaction in recipients receiving a vaccine made from *Aedes* or *Toxorhynchites* cell cultures makes these substrates unsuitable for these purposes without the removal of allergens from these vaccines. Preliminary experiments indicate that the allergen is not virion-associated and is probably a cellular breakdown or secretory product. Attenuation with optimal virus yields from C6/36 cell cultures followed by inoculation of acceptable mammalian cells for the final vaccine product is another alternative. Both of these approaches are being explored for future DEN vaccine preparation.

ACKNOWLEDGMENTS

The authors would like to thank Dr. Akira Igarashi for the C6/36 clone of *Aedes albopictus*, Dr. Robert Tesh for the TA-9 and TA-42 cell lines, and Dr. Goro Kuno for the TRA-284 cell line. We would also like to thank Dr. Michael Barile for performing mycoplasma assays, Dr. Julie Milstein for reverse transcriptase assays, Dr. Jean Adams for electron microscopy, Dr. Dennis Knudson for isoenzyme analyses, and Dr. Dennis Brown for fusion of C6/36 cells with other invertebrate and mammalian cells. Dr. Artie Shelton and Dr. Richard Summers were instrumental in performing the human and monkey skin tests. We would also like to thank Dr. D. R. Dubois for providing laboratory and cell culture support and expert guidance.

REFERENCES

1. Singh, K. R. P., Cell cultures derived from larvae of *Aedes albopictus* (Skuse) and *Aedes aegypti* (L), *Curr. Sci.*, 36, 506, 1967.
2. Singh, K. R. P. and Paul, S. D., Multiplication of arboviruses in cell lines from *Aedes albopictus* and *Aedes aegypti, Curr. Sci.*, 37, 65, 1968.
3. Paul, S. D. and Singh, K. R. P., Comparative sensitivity of mosquito cell lines, Vero cell line, and infant mice to infection with arboviruses, *Curr. Sci.*, 38, 241, 1969.
4. Singh, K. R. P. and Paul, S. D., Isolation of dengue viruses in *Aedes albopictus* cell cultures, *Bull. W.H.O.*, 40, 982, 1969.
5. Suitor, E. C., Jr. and Paul, F. J., Syncytia formation of mosquito cell cultures mediated by type 2 dengue virus, *Virology*, 38, 482, 1969.
6. Stevens, T. M., Arbovirus replication in mosquito cell lines (Singh) grown in monolayer or suspension culture, *Proc. Soc. Exp. Biol. Med.*, 134, 356, 1970.
7. Igarashi, A., Isolation of a Singh's *Aedes albopictus* cell clone sensitive to dengue and chikungunya viruses, *J. Gen. Virol.*, 40, 531, 1978.
8. Igarashi, A., Characteristics of *Aedes albopictus* cells persistently infected with dengue viruses, *Nature*, 280, 690, 1979.
9. Dubois, D. R., Unpublished data, 1983.
10. Eckels, K. H., Unpublished date, 1983.
11. Code of Federal Regulations (U.S.A.), Food and Drugs, Title 21, Parts 600-799, April, 1983.
12. Scott, R. McN., Shelton, A. L., Eckels, K. H., Bancroft, W. H., Summers, R. J., and Russell, P. K., Human hypersensitivity to a sham vaccine prepared from mosquito-cell culture fluids, *J. Allergy Clin. Immunol.*, 74(6), 808, 1984.
13. Theiler, M., The virus, in *Yellow Fever,* Strode, G. K., Ed., McGraw-Hill, New York, 1963, 39.
14. Petermans, J. and Huygelen, C., The attenuation of rubella virus by serial passage in primary rabbit kidney cell cultures. I. Growth characteristics *in vitro* and production of experimental vaccines at different passage levels, *Arch. Gesamte Virusforsch.*, 21, 133, 1967.
15. Holland, J., Spindler, K., Horodyski, F., Grabau, E., Nichol, S., and Vande Pol, S., Rapid evolution of RNA genomes, *Science*, 215, 1577, 1982.
16. Eckels, K. H., Brandt, W. E., Harrison, V. R., McCown, J. M., and Russell, P. K., Isolation of a temperature-sensitive dengue-2 virus under conditions suitable for vaccine development, *Infect. Immun.*, 14, 1221, 1976.
17. Eckels, K. H., Harrison, V. R., Summers, P. L., and Russell, P. K., Dengue-2 vaccine: preparation from a small-plaque virus clone, *Infect. Immun.*, 27, 175, 1980.
18. Harrison, V. R., Eckels, K. H., Sagartz, J. W., and Russell, P. K., Virulence and immunogenicity of a temperature-sensitive dengue-2 virus in lower primates, *Infect. Immun.*, 18, 151, 1977.
19. Bancroft, W. H., Top, F. H., Jr., Eckels, K. H., Anderson, J. H., Jr., McCown, J. M., and Russell, P. K., Dengue-2 vaccine: virological, immunological, and clinical responses of six yellow fever-immune recipients, *Infect. Immun.*, 31, 698, 1981.
20. Scott, R. McN., Eckels, K. H., Bancroft, W. H., Summers, P. L., McCown, J. M., Anderson, J. H., and Russell, P. K., Dengue-2 vaccine: dose response in volunteers in relation to yellow fever immune status, *J. Infect. Dis.*, 148, 1055, 1983.
21. Halstead, S. B., Diwan, A. R., Marchette, N. J., Palumbo, N. E., and Srisukonth, L., Selection of attenuated dengue-4 viruses by serial passage in primary kidney cells. I. Attributes of uncloned virus at different passage levels, *Am. J. Trop. Med. Hyg.*, 33, 684, 1984.

Replicative Dynamics of Arboviruses in Arthropod Cell Systems

Chapter 13

COMPARATIVE GROWTH OF ARBOVIRUSES IN CELL LINES DERIVED FROM *AEDES* AND *ANOPHELES* MOSQUITOES AND FROM THE TICK *BOOPHILUS MICROPLUS*

C. J. Leake

TABLE OF CONTENTS

I.   Introduction 26

II.   Cell Lines and Their Infection with Arboviruses 26

III.   Aspects of Arbovirus Replication 27
         A.   Minimum Infectious Dose Experiments 27
         B.   Kinetics of Arbovirus Growth in Cell Cultures 27
         C.   Persistent Infections 39
         D.   Cytopathic Effects, Immunofluorescence, and
                 Immunocytochemistry 40

IV.   Conclusion 40

Acknowledgments 41

References 41

I. INTRODUCTION

The techniques for initiating cell cultures from mosquitoes and ticks have been developed to the point where it is now relatively easy to establish new cell lines, particularly from mosquitoes.[1] Characterization and evaluation of cell lines for their susceptibility to infection with arboviruses is, however, a time-consuming process and it is perhaps not too surprising that relatively few of the established cell lines have been studied in sufficient detail to allow their wide use in arbovirus research. In this chapter work will be described which, to my knowledge, is still unique: the study, in parallel, of the growth of a large number of arboviruses in five continuous mosquito cell lines. This represents the only comprehensive study of *Anopheline* cell lines to have been carried out. Subsequently, we established a new tick cell line from the cattle tick *Boophilus microplus* and described preliminary studies in 1980.[2] The full results of virus growth studies in these cells described here allow a number of interesting comparisons to be made between arbovirus infected tick and mosquito cells in vitro.

II. CELL LINES AND THEIR INFECTION WITH ARBOVIRUSES

The cell lines used in these studies were

1. *Aedes aegypti* LSTM-AA-20A (mosquito) cell line[3] used between subcultures 231 and 373.
2. *A. malayensis* LSTM-AM-60 (mosquito) cell line[4] used between subcultures 32 and 180.
3. *A. pseudoscutellaris* LSTM-AP-61 (mosquito) cell line[4] used between subcultures 35 and 176.
4. *Anopheles stephensi* LSTM-AS-43 (mosquito) cell line[5] used between subcultures 199 and 335.
5. *A. gambiae* LSTM-AG-55 (mosquito) cell line[6] used between subcultures 128 and 264.
6. *Boophilus microplus* LSTM-BM-270 (tick) cell line[2,7] used between subcultures 29 and 66.

All of the lines were maintained at 28°C as originally described, except that proteolytic enzymes were not used routinely. The BM-270 cells were initiated at 28°C in Hanks HLH medium,[1] but subsequently a more vigorous subline adapted to grow in L-15 medium supplemented with 10% tryptose phosphate broth and 15% fetal bovine serum (FBS) at the higher temperature of 32°C was used. For virus infection the cells were grown to confluency in 1 oz glass prescription bottles or in 25 cm² plastic flasks in L-15 medium and infected and harvested as described earlier.[8] The only variation from these methods was in experiments with the BM-270 cells where infecting virus suspensions were allowed to adsorb for 1 hr at 32°C compared to 3 hr at 28°C for the mosquito cells.

There were 26 viruses originally prepared as suckling mouse brain (SMB) suspensions used (Table 1). Titers ($\log_{10}$PFU/mℓ) were determined by plaque assay using BHK-21, PS and Vero (mammalian) cell lines, and the *Xenopus laevis* (XTC-2) (lower vertebrate) cell line,[8,9] except for Middelburg and Ganjam viruses, where no clear plaques were formed and titers ($\log_{10}$TCD 50/mℓ) were determined by assessing the extent of cytopathic effects (CPE) produced in XTC-2 cells.

Table 1
VIRUS STRAINS

Virus	Group	Abbrev.	Strain	Passage	Titer/ml	Assay
Chikungunya	A	CHIK	E103	4	6.0 PFU	XTC-2
Middelburg	A	MID	AR-749	16	7.8 TCD50	XTC-2
Ndumu	A	NDU	AR-2211	9	7.9 PFU	XTC-2
O'Nyong-Nyong	A	ONN	AHERO	12	6.7 PFU	BHK-21
Semliki Forest	A	SF	DB478MH	13	8.0 PFU	VERO
Sindbis	A	SIN	AR-339	1	7.2 PFU	XTC-2
Mayaro	A	MAY	TR-4675	11	8.4 PFU	XTC-2
Dengue-2	B	DEN-2	N. Guinea	34	6.7 PFU	PS
Japanese encephalitis	B	JE	Nakayama	1	8.9 PFU	PS
Langat	B	LGT	TP-64	6	8.9 PFU	PS
Louping Ill	B	LI	369T2	4	8.2 PFU	PS
West Nile	B	WN	E101	10	7.4 PFU	VERO
Yellow Fever	B	YF	17D	24	7.4 PFU	VERO
Zika	B	ZIKA	MR-766	4	7.6 PFU	PS
Anopheles A	BUN	ANA	Proto	17	6.8 PFU	VERO
Bunyamwera	BUN	BUN	Original	20	8.9 PFU	VERO
Bwamba	BUN	BWA	M-459	1	7.1 PFU	VERO
California enceph- alitis	CAL	CE	BFS-283	12	6.9 PFU	VERO
Ganjam	NSD	GAN	G-619	10	5.2 TCD50	XTC-2
Germiston	BUN	GER	AR-1050	12	9.2 PFU	VERO
Sandfly fever	PHL	SFS	Sicilian	1	6.2 PFU	PS
Chandipura	VSV	CHP	I-653514	1	9.2 PFU	VERO
Piry	VSV	PIRY	An-24232	1	9.0 PFU	VERO
Keterah	UCL	KET	P6-1361	8	6.8 PFU	XTC-2
Quaranfil	QRF	QRF	AR-1095	23	6.9 PFU	XTC-2
Zirqa	HUG	ZIR	A-2070-1	4	6.1 PFU	XTC-2

III. ASPECTS OF ARBOVIRUS REPLICATION

A. Minimum Infectious Dose Experiments

The sensitivity of the mosquito cells to virus infection was compared by inoculating serial decimal dilutions of virus suspensions and titrating harvests 3 and 10 days later to determine if the cells were infected. In experiments with eight viruses (Table 2) it became immediately apparent that there were considerable differences in the sensitivity of these cell lines. The AA-20A cells were generally more sensitive to some of the viruses tested than both of the anopheline cell lines, with the AS-43 cells being the least sensitive. In marked contrast, the AM-60 and AP-61 cells were very similar to each other, being much more sensitive than the other mosquito cell lines or for that matter the control assay systems. This argued that these cells might have applications as isolation systems.

B. Kinetics of Arbovirus Growth in Cell Cultures

Initial studies performed with the three alphaviruses SIN, SF (Figure 1), and ONN (Figure 3), and the flaviviruses JE (Figure 3) and WN (Figure 4) revealed that the AM-60 and AP-61 cells were not only more sensitive than the other mosquito cell lines, but also supported virus replication faster and to a higher titer. Experiments were then carried out to compare the susceptibility of the mosquito cell lines to an additional 18 viruses from several antigenic groups. The working assumptions proved correct in that the AM-60 and AP-61 cells continued to prove superior to the other cell lines over this range. Both lines supported the replication of 18/23 of the viruses tested (Figures 1 to 9) and, additionally, low levels of SFS virus (Figure 6) were detected for several days,

Table 2

MINIMUM INFECTIOUS DOSE EXPERIMENTS[a]

Virus	Dilution Log$_{10}$	AA-20A	AM-60	AP-61	AS-43	AG-55[b]
SIN	-1	-	+	+	-	+
	-5	-	+	+	-	+
	-6	-	+	+	-	-
	-9	-	+	+	-	-
SF	-1	+	+	+	-	+
	-5	+	+	+	-	+
	-6	-	+	+	-	+
	-7	-	+	+	-	-
	-9	-	+	+	-	-
CHIK	-3	+	+	+	-	+
	-5	+	+	+	-	-
	-6	-	+	+	-	-
	-7	-	+	+	-	-
ONN	-1	-	+	+	+	+
	-5	-	+	+	-	+
	-6	-	+	+	-	-
	-7	-	-	+	-	-
WN	-6	+	+	+	+	+
	-7	-	+	+	-	-
	-8	-	+	+	-	-
	-9	-	+	+	-	-
YF	-2	+	+	+	-	-
	-5	+	+	+	-	-
	-6	+	+	+	-	-
	-7	-	+	+	-	-
	-8	-	+	+	-	-
BUN	-5	+	+	+	+	+
	-6	+	+	+	-	-
	-7	-	+	+	-	-
	-8	-	-	-	-	-
CE	-4	+	+	+	+	+
	-5	+	+	+	+	-
	-6	+	+	+	+	-
	-7	-	+	-	-	-

[a] Refer to Table 1 for stock virus titers.

[b] Abbreviations for cell lines: AA-20A is *Aedes aegypti* LSTM-AA-20A; AM-60 is *A. malayensis* LSTM-AM-60; AP-61 is *A. pseudoscutellaris* LSTM-AP-61; AS-43 is *Anopheles stephensi* LSTM-AS-43; and AG-55 is *A. gambiae* LSTM-AG-55.

which may or may not have indicated replication. In connection with this, more recent work with another *Phlebovirus*, Toscana, has indicated that the AP-61 cells will support slow replication of this virus as measured both by plaque assay and indirect immunofluorescence.[10] The next best cell line was the AA-20A line supporting replication of 14/23 viruses tested. Replication of DEN-2 virus was equivocal as low levels of DEN-2 virus were only detected late in the experiments on days 10 and 14 (Figure 4). The AG-55 cell line supported the replication of 11/23 viruses, where BUN (Figure 5)

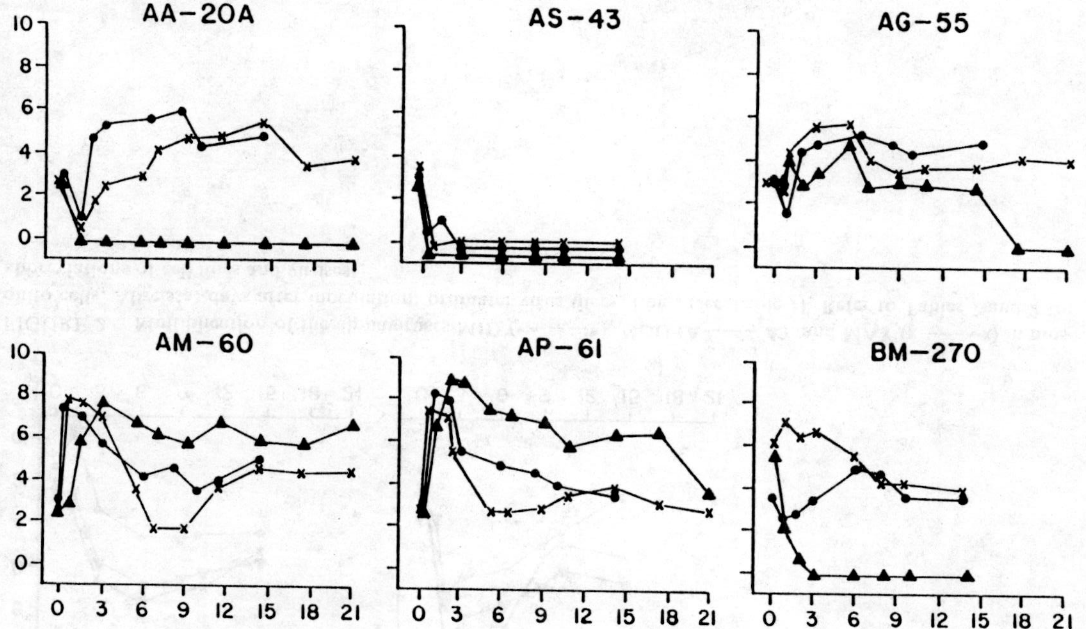

FIGURE 1.   Multiplication of the alphaviruses CHIK (•———•), SIN (▲———▲), and SF (x———x) in mosquito and tick cells. Abscissa; days after inoculation; ordinate: virus titers, Log_{10} (see Table 1). Refer to Tables 1 and 2 for abbreviations of cell lines and viruses.

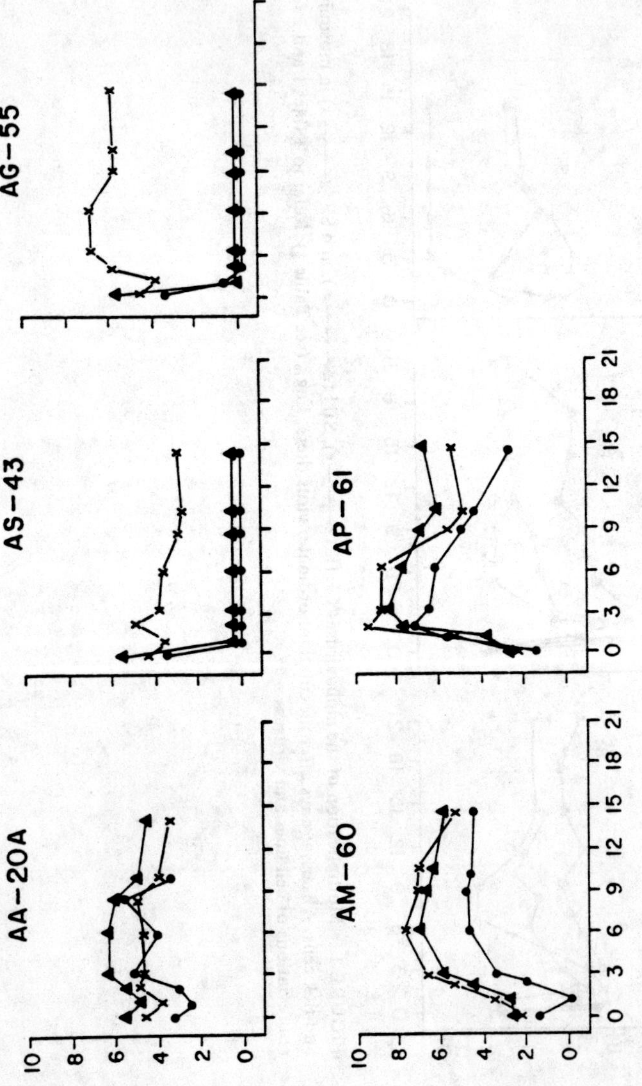

FIGURE 2. Multiplication of the alphaviruses MID (•——•), NDU (▲——▲), and MAY (x——x) in mosquito cells. Abscissa: days after inoculation; ordinate: virus titers, Log$_{10}$ (see Table 1). Refer to Tables 1 and 2 for abbreviations of cell lines and viruses.

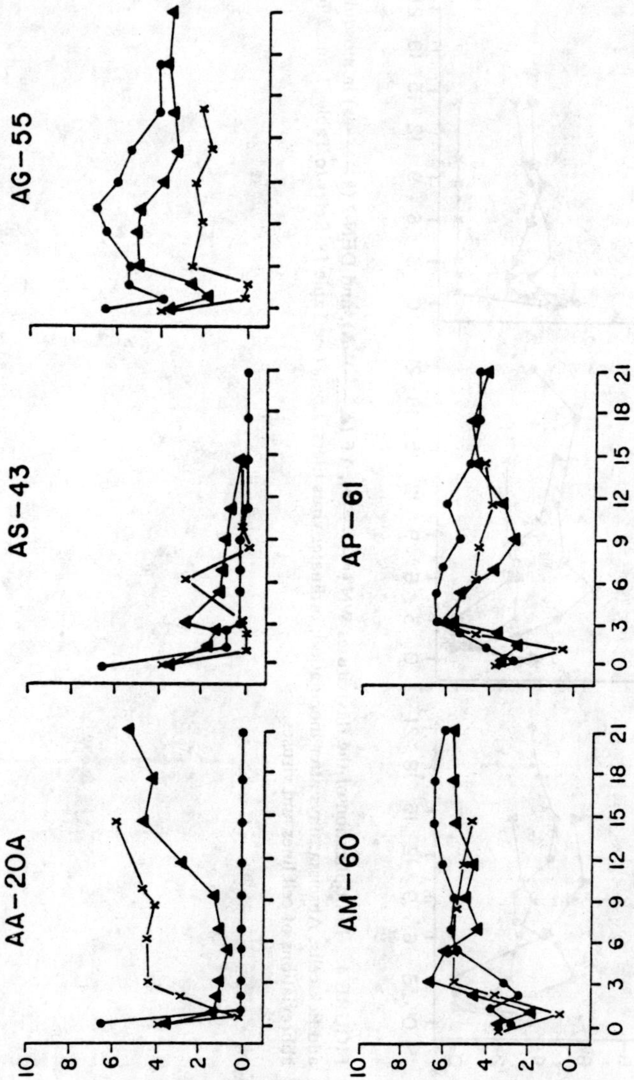

FIGURE 3. Multiplication of the *Alphavirus* (ONN (•———•), and the flaviviruses JE (▲———▲) and ZIKA (×———×) in mosquito cells. Abscissa: days after inoculation; ordinate: virus titers, Log_{10} (see Table 1). Refer to Tables 1 and 2 for abbreviations of cell lines and viruses.

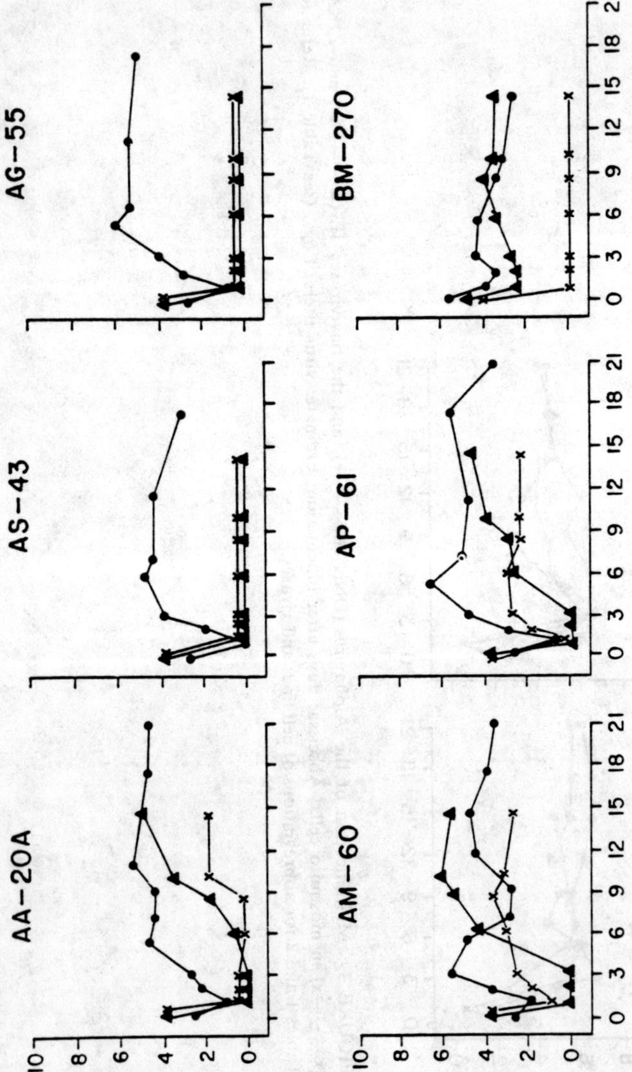

FIGURE 4.   Multiplication of the flaviviruses WN (•——•), YF (▲——▲), and DEN-2 (x——x) in mosquito and tick cells. Abscissa: days after inoculation; ordinate: virus titers, Log_{10} (see Table 1). Refer to Tables 1 and 2 for abbreviations of cell lines and viruses.

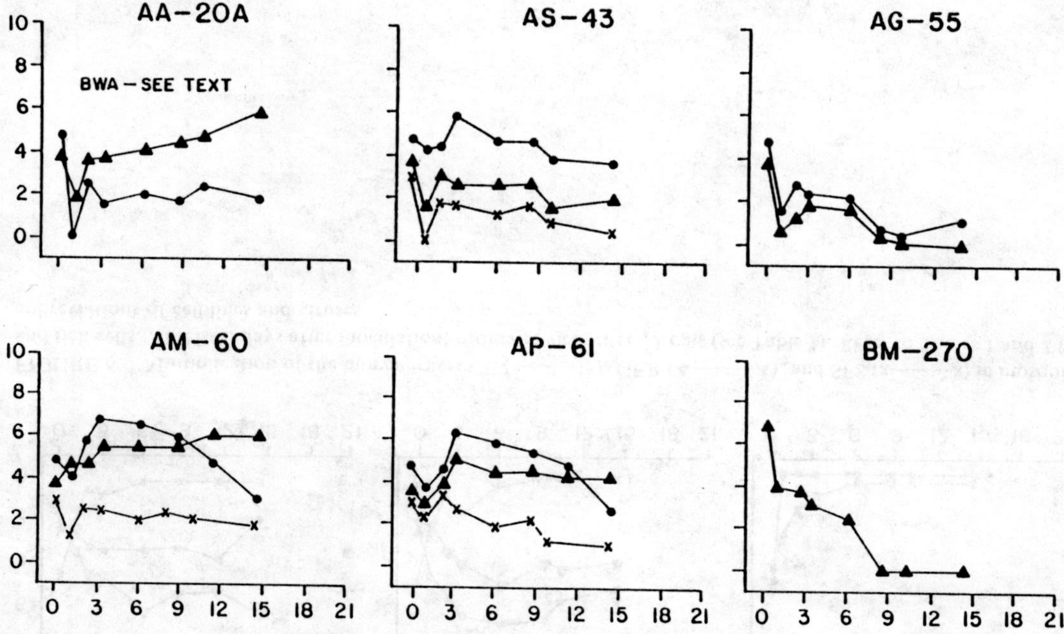

FIGURE 5. Multiplication of the bunyaviruses ANA (•————•), BUN (▲————▲), and BWA (x————x) in mosquito and tick cells. Abscissa: days after inoculation; ordinate: virus titers, Log_{10} (see Table 1). Refer to Tables 1 and 2 for abbreviations of cell lines and viruses.

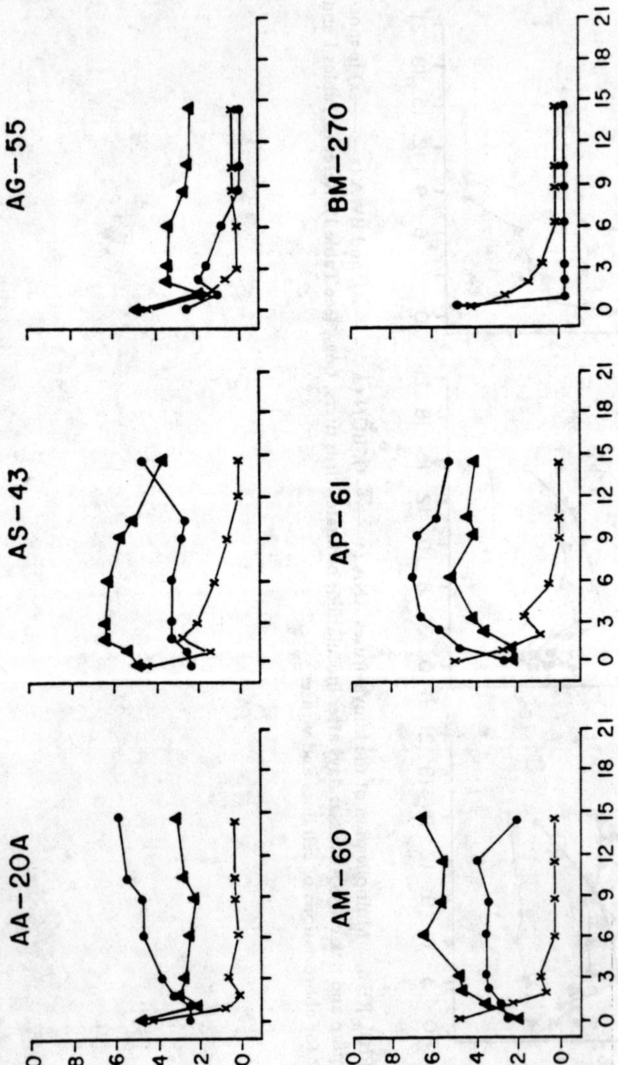

FIGURE 6.   Multiplication of the bunyaviruses CE (•———•), GER (▲———▲), and SFS (x———x) in mosquito and tick cells. Abscissa: days after inoculation; ordinate: virus titers, Log$_{10}$ (see Table 1). Refer to Tables 1 and 2 for abbreviations of cell lines and viruses.

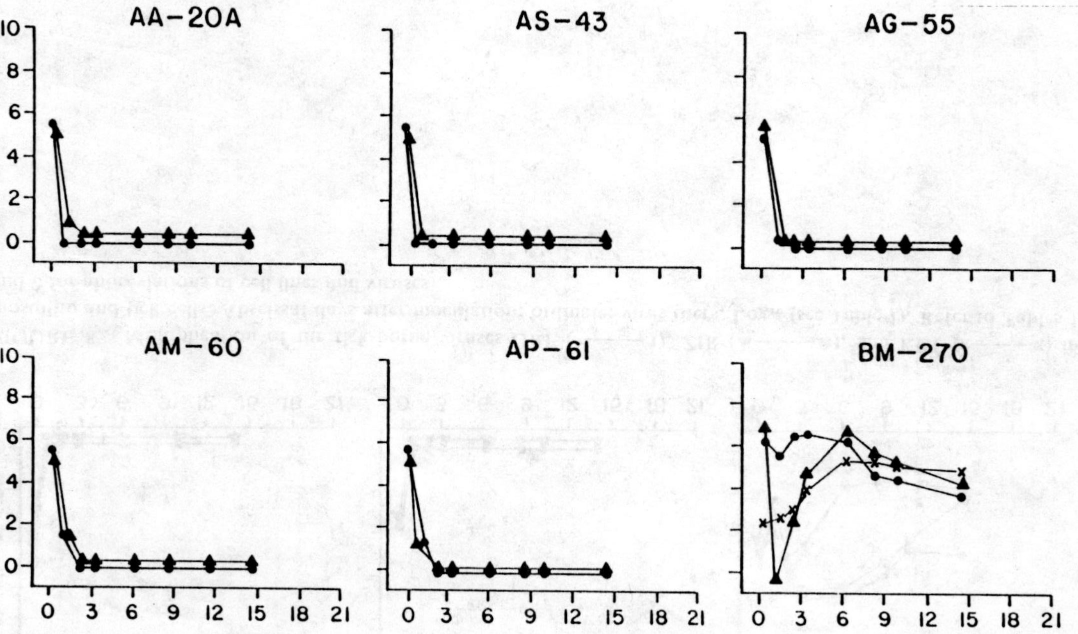

FIGURE 7. Multiplication of the tick-borne flaviviruses LI (•————•), and LGT (▲————▲), and the bunyavirus GJM (x————x) in mosquito and tick cells. Abscissa: days after inoculation; ordinate: virus titers, Log$_{10}$ (see Table 1). Refer to Tables 1 and 2 for abbreviations of cell lines and viruses.

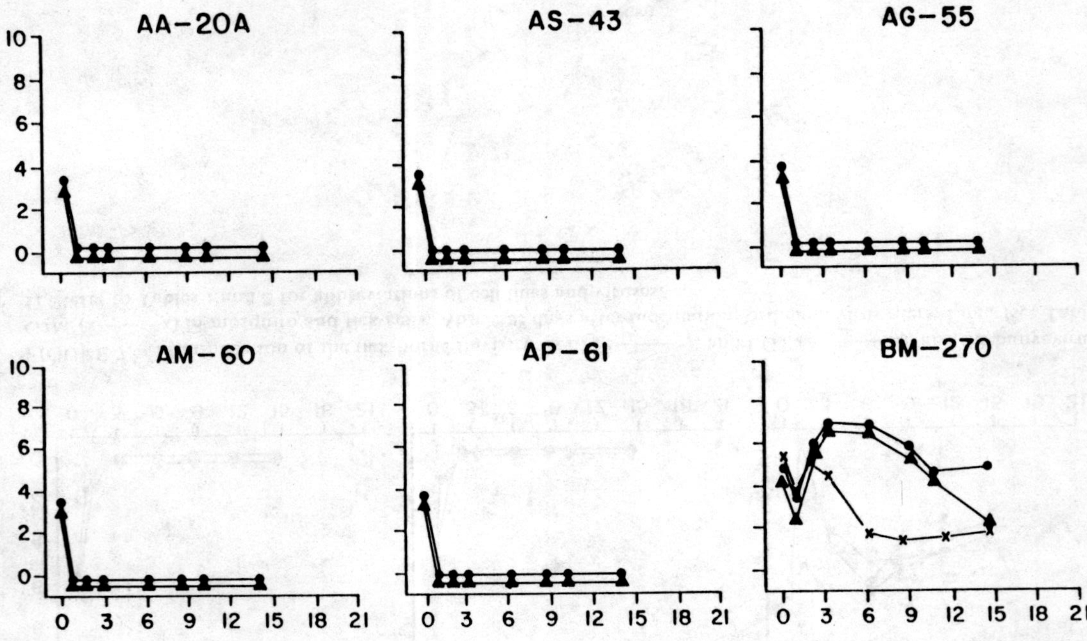

FIGURE 8.   Multiplication of the tick-borne viruses QRF (•———•), ZIR (▲———▲), and KET (x———x) in mosquito and tick cells. Abscissa: days after inoculation; ordinate: virus titers, Log_{10} (see Table 1). Refer to Tables 1 and 2 for abbreviations of cell lines and viruses.

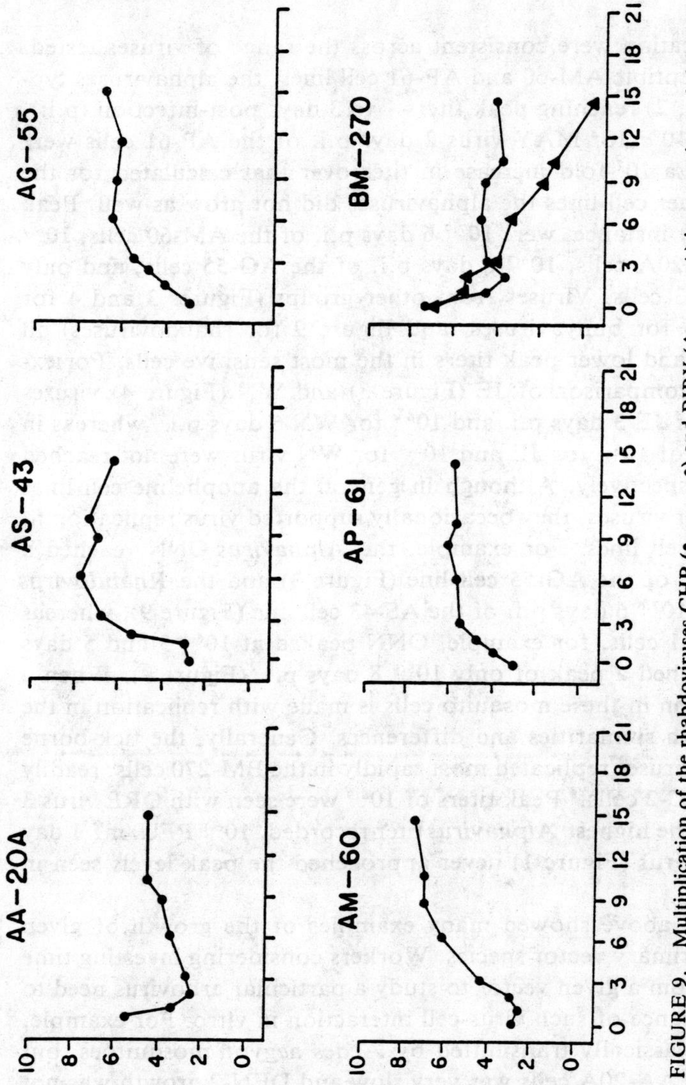

FIGURE 9. Multiplication of the rhabdoviruses CHP (•———•) and PIRY (▲———▲) in mosquito and tick cells. Abscissa: days after inoculation; ordinate: virus titers, Log_{10} (see Table 1). Refer to Tables 1 and 2 for abbreviations of cell lines and viruses.

and CE (Figure 6) viruses were detected at low levels for several days. Finally, the AS-43 cells supported the replication of only 8/23 viruses tested; falling or very low titers were observed with CHIK (Figure 1), ONN, JE, ZIKA (Figure 3), and SFS (Figure 6) viruses.

An important finding was the consistent failure of all the tick-borne viruses, LI and LGT (Figure 7) and QRF and ZIR (Figure 8), to grow in any of the mosquito cell lines. In comparison with this observation, the BM-270 cells supported replication of 11/18 viruses tested including all the tick-borne viruses tested as well as several mosquito-borne viruses. These results were consistent with our earlier studies on other tick cell lines.[2,11]

The kinetics of virus replication were consistent across the range of viruses tested. In general, in the highly susceptible AM-60 and AP-61 cell lines, the alphaviruses typically grew rapidly (Figures 1, 2) reaching peak titers 1 to 3 days post-infection (p.i.). Titers (PFU/mℓ) as high as $10^{9.2}$ for MAY virus 2 days p.i. of the AP-61 cells were seen (Figure 2), representing a 10^7-fold increase in titer over that calculated for the inoculum. In some of the other cell lines the alphaviruses did not grow as well. Peak titers for MAY (Figure 2), for instance, were $10^{7.7}$ 6 days p.i. of the AM-60 cells, $10^{6.1}$ 3 and 6 days p.i. of the AA-20A cells, $10^{6.7}$ 6 days p.i. of the AG-55 cells, and only $10^{4.8}$ 2 days p.i. of the AS-43 cells. Viruses from other groups (Figures 3 and 4 for flaviviruses; Figures 5 and 6 for bunyaviruses; and Figure 9 for rhabdoviruses) all tended to grow more slowly and lower peak titers in the most sensitive cells. For example in the AP-61 cells, a comparison of JE (Figure 3) and WN (Figure 4) viruses showed peak titers of $10^{5.7}$ for JE 3 days p.i. and $10^{6.4}$ for WN 6 days p.i., whereas in the AA-20A cells peak titers of $10^{5.1}$ for JE and $10^{5.4}$ for WN virus were not reached until 21 and 11 days p.i., respectively. Although in general the anopheline cell lines supported replication of fewer viruses, they occasionally supported virus replication to higher titers than the other cell lines. For example, the *Alphavirus* ONN reached a peak titer of $10^{6.4}$ 7 days p.i. of the AG-55 cell line (Figure 3) and the *Rhabdovirus* CHP reached a peak titer of $10^{7.6}$ 6 days p.i. of the AS-43 cell line (Figure 9), whereas in the broadly sensitive AP-61 cells, for example, ONN peaked at $10^{6.1}$ 3 and 5 days p.i. (Figure 3) and CHP reached a peak of only $10^{5.9}$ 8 days p.i. (Figure 9). When a comparison of virus replication in these mosquito cells is made with replication in the BM-270 cells, we can see both similarities and differences. Generally, the tick-borne flaviviruses and unclassified viruses replicated most rapidly in the BM-270 cells, readily exceeding their growth in XTC-2 cells.[8] Peak titers of $10^{6.9}$ were seen with QRF virus 3 days p.i. (Figure 8), whereas the highest *Alphavirus* titer recorded, $10^{6.8}$ PFU/mℓ 1 day p.i. with a high dose of SF virus (Figure 1) never approached the peak levels seen in mosquito cells.

The experiments described above showed many examples of the growth of given arboviruses in cells of their primary vector species. Workers considering investing time in establishing cell cultures from a given vector to study a particular arbovirus need to consider carefully the significance of such virus-cell interaction in vitro. For example, DEN-2 and YF viruses are classically transmitted by *Aedes aegypti* mosquitoes, but replication of YF virus in the AA-20A cells was very slow and DEN-2 growth was not conclusively demonstrated (Figure 4). It was also interesting that the tick cell lines supported replication of several of the viruses tested, including YF (Figure 4). Although YF has been isolated from both eggs and larvae of the tick *Amblyomma variegatum*,[12] ticks have not generally been considered important in the transmission of the YF virus. Similarly, anopheline mosquitoes, while principally vectors of malaria, have not really been considered important arbovirus vectors. It should be remembered, however, that *Anopheles gambiae* transmitted the *Alphavirus* ONN very efficiently in the huge African epidemic in the late 1950s. In vitro ONN grew very well in the AG-55

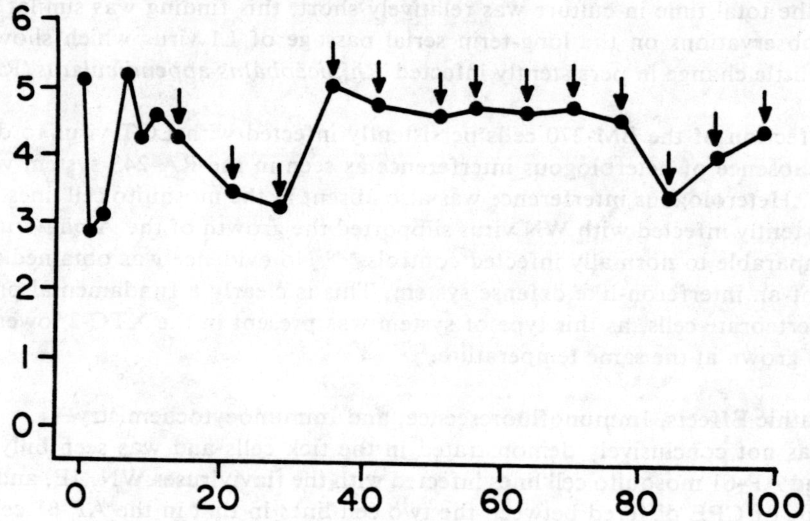

FIGURE 10.   Persistent infection of *Boophilus microplus* (BM-270) cells with the *Flavivirus* Langat. Abscissa: days after inoculation; ordinate: virus titers, Log_{10} (see Table 1). ↓ indicates the cells that were subcultured.

cells (Figure 3). JE virus also grew well in these cells (Figure 3), and indeed several species of anopheline mosquitoes have been implicated in human transmission of JE virus. Greater attention should be placed on anopheline mosquitoes by arbovirologists in both field and laboratory vector-competence experiments. The experience gathered in these studies indicates that the best course for in vitro studies of arboviruses is to use one of the established cell lines or to select virus-sensitive clones from these lines as Igarashi[23] has done, rather than to go through the laborious procedure of establishing and characterizing new cell lines which may have limited potential.

Overall, the AP-61 cell line was selected as the system most suitable for evaluation of primary virus isolation. Not only was it sensitive and broadly susceptible compared with most of the other cell lines, but in comparison with the AM-60 cell line which had similar sensitivity and susceptibility, the AP-61 cell line had generally superior handling characteristics. Some applications of the AP-61 cell line for virus isolation are described below and elsewhere in this book.

C. Persistent Infections

It will be noted from the growth curves that in most cases the cells were still productively infected at the termination of the experiments and, in several cases, cells were easily subcultured for extended periods. A culture of AM-60 cells infected with the WN virus was still productively infected over 2 years later and a similar phenomenon was seen with the *Flavivirus* LGT in the BM-270 cells, where it was possible to subculture the cells 12 times over 98 days before the culture was lost due to contamination (Figure 10). Although persistently infected cultures from all of the cell lines were not examined, it seems likely that these systems could readily have been established using the AA-20A, AG-55, and AS-43 cells provided that virus growth was initially demonstrated. Alterations in the plaque morphology of SF, SIN, CHIK, and WN viruses with a selection for small plaque variants were noted in persistently infected AM-60 and AP-61 cultures. This selection occurred at varying rates and when the growth of BWA virus was examined in the AA-20A cells, a small plaque variant emerged immediately which produced such tiny plaques that an accurate count was not possible. No apparent alterations were seen in the plaque morphology of LGT virus in the BM-270 cells and although the total time in culture was relatively short, this finding was similar to our previous observations on the long-term serial passage of LI virus which showed remarkably little change in persistently infected *Rhipicephalus appendiculatus* (RA-243) cells.[2]

Superinfection of the BM-270 cells persistently infected with LGT virus to demonstrate the absence of heterologous interference as seen in the RA-243 system was not attempted. Heterologous interference was also absent in the mosquito cell lines, where cells persistently infected with WN virus supported the growth of the *Alphavirus* SF to levels comparable to normally infected controls.[8,13] No evidence was obtained for the presence of an interferon-like defense system. This is clearly a fundamental property of the invertebrate cells, as this type of system was present in the XTC-2 lower vertebrate cells grown at the same temperature.[8]

D. Cytopathic Effects, Immunofluorescence, and Immunocytochemistry

CPE was not conclusively demonstrated in the tick cells and was seen only in the AM-60 and AP-61 mosquito cell lines infected with the flaviviruses WN, JE, and DEN-2. The type of CPE differed between the two cell lines in that in the AP-61 cells syncytial formation was seen, whereas in the AM-60 cells CPE was mainly a focal cell-death. In other studies we have also seen focal CPE in the AP-61 cells infected with wild strains of the YF virus as well as syncytial CPE with primary isolates of the alphaviruses Sindbis and Getah.[13,14] Plaque production in mosquito cell monolayers under semisolid overlays has been inconsistent, but has been used in a plaque neutralization test for wild strains of the YF virus in Trinidad.[15] Plaque formation under agarose overlays has been more useful,[13] but generally has not been widely used. As cytopathology in these mosquito cells is a phenomenon which seems to vary considerably depending on a variety of factors including cell passage level and cultural conditions, we usually recommend, for example, that for dengue virus isolation based on CPE, the AP-61 cells should not be used above subculture 50. From a practical point of view this is very restricting, particularly as the cells are still able to support the growth of the viruses at higher passage levels, although CPE may be absent or more difficult to determine. The recent availability of type-specific monoclonal antibodies to the dengue viruses[16] and type-specific YF, JE, and *Flavivirus* group-specific monoclonal antibodies[17,18] have provided new tools for the use of these mosquito cells for rapid diagnosis and typing of virus isolates by the indirect immunofluorescence technique. This has been used very successfully in Thailand for a comparison of the isolation of dengue viruses from Dengue Hemorrhagic fever patients[19] and JE virus strains from encephalitis patients, animal sera, and mosquito pools.[20,21] Although the IFA technique is rapid and quite sensitive, it has the drawbacks of requiring an expensive fluorescence microscope and produces nonpermanent preparations which are subject to UV photobleaching effects. To overcome these problems we have begun to explore the use of immunoperoxidase staining of virus-infected cells.[22] Using monoclonal antibodies and avidin-biotin amplified immunoperoxidase staining which offer greatly enhanced sensitivity over available IFA techniques, we have successfully localized the JE and DEN-2 virus antigen in infected AP-61 cell cultures using an ordinary light microscope. In addition, the preparations may be permanently mounted for archival purposes and repeated examination.

IV. CONCLUSION

In a very short time a limited number of mosquito cell lines has been developed from laboratory curiosities to widely distributed and valuable tools for the study of arboviruses. Clearly the field is still developing and this can be illustrated by the availability of cloned mosquito cells[23] and mosquito cells adapted to serum-free media.[24,25] In contrast, tick cell culture is in imminent danger of becoming a dying art and in view of the human and veterinary viruses transmitted by ticks this is a dangerous situation. In the immediate future, however, it seems unlikely that the type of studies described in this chapter will need to be repeated. The tools are available and the challenge now lies in using these tools to answer many questions in the life-history and biology of the arboviruses.

ACKNOWLEDGMENTS

The work described was performed under successive grants from the Medical Research Council of Great Britain and from the Wellcome Trust. The work would have been impossible without the encouragement, advice, and assistance of my colleagues, Prof. M. G. R. Varma and Dr. Mary Pudney. Much of the work formed part of a Ph.D. thesis submitted to the University of London. I am currently supported by the Wellcome Trust at the U.S. Component of the Armed Forces Research Institute of Medical Sciences in Bangkok, and I am grateful to my colleagues in the Department of Virology for their critical comments during preparation of this manuscript.

REFERENCES

1. **Pudney, M., Varma, M. G. R., and Leake, C. J.**, I. Establishment of cell lines from larvae of culicine (*Aedes* species) and anopheline mosquitoes. II. Establishment of cell lines from ixoidid ticks, *Tissue Cult. Assoc. Man.*, 5, 997, 1979.
2. **Leake, C. J., Pudney, M., and Varma, M. G. R.**, Studies on arboviruses in established tick cell lines, in *Invertebrate Systems In Vitro*, Kurstak, E., Maramorosch, K., and Dübendorfer, A., Eds., Elsevier/North Holland, Amsterdam, 1980, 327.
3. **Varma, M. G. R. and Pudney, M.**, The growth and serial passage of cell lines from *Aedes aegypti* (*L.*) larvae in different media, *J. Med. Entomol.*, 6, 432, 1969.
4. **Varma, M. G. R., Pudney, M., and Leake, C. J.**, Cell lines from larvae of *Aedes (Stegomyia) malayensis* (Colless) and *Aedes (S) pseudoscutellaris* (Theobald) and their infection with some arboviruses, *Trans. R. Soc. Trop. Med. Hyg.*, 68, 374, 1974.
5. **Pudney, M. and Varma, M. G. R.**, *Anopheles stephensi var mysorensis*: establishment of a larval cell line (Mos 43), *Exp. Parasitol.*, 29, 7, 1971.
6. **Marhoul, Z. and Pudney, M.**, A mosquito cell line (Mos 55) from *Anopheles gambiae* larvae, *Trans. R. Soc. Trop. Med. Hyg.*, 66, 183, 1972.
7. **Pudney, M. and Leake, C. J.**, Unpublished data, 1978.
8. **Leake, C. J.**, Comparative Studies on the Infection of Invertebrate and Vertebrate Cell Lines with Some Arboviruses, Unpublished Ph.D. thesis, University of London, 1977.
9. **Leake, C. J., Varma, M. G. R., and Pudney, M.**, Cytopathic effect and plaque formation by some arboviruses in a continuous cell line (XTC-2) from the toad *Xenopus laevis*, *J. Gen. Virol.*, 35, 335, 1977.
10. **Niccoletti, L. and Leake, C. J.**, Unpublished data, 1982.
11. **Pudney, M., Varma, M. G. R., and Leake, C. J.**, The growth of some arboviruses in tick cell lines, in *Tick-Borne Diseases and Their Vectors*, Wilde, J. K. H., Ed., University of Edinburgh, 1978, 490.
12. **Saluzzo, J. F., Herve, J. P., Saluan, J. J., Germain, M., Cornet, J. P., Camicas, J. L., Heme, G., and Robin, Y.**, Caractéristiques des souches du virus de la fièvre jaune isolées des oeufs et des larves d'une tique *Amblyomma variegatum* récoltée sur le bétail à Bangui (Centrafrique), *Ann. Virol. (Inst. Pasteur)*, 131E, 155, 1980.
13. **Pudney, M., Leake, C. J., and Buckley, S. M.**, Replication of arboviruses in arthropod *in vitro* systems: an overview, in *Invertebrate Cell Culture Applications*, Maramorosch, K. and Mitsuhashi, J., Eds., Academic Press, New York, 1982, 159.
14. **Varma, M. G. R., Pudney, M., Leake, C. J., and Peralta, P. H.**, Isolations in a mosquito (*Aedes pseudoscutellaris*) cell line (Mos 61) of yellow fever virus strains from original field material, *Intervirology*, 6, 50, 1976.

15. Leake, C. J., Unpublished data, 1979.
16. Henchal, E., McCown, J. M., Seguin, M. C., Gentry, M. K., and Brandt, W. E., Rapid identification of dengue virus isolates by using monoclonal antibodies in an indirect immunofluorescence assay, *Am. J. Trop. Med. Hyg.*, 32, 164, 1983.
17. Burke, D. S., Personal communications, 1983, 1984.
18. Gould, E. A., Personal communications, 1982 to 1984.
19. Leake, C. J., Nisalak, A., and Burke, D. S., Comparative isolation of dengue viruses by mosquito inoculation and on three mosquito cell lines, in *Proceedings of International Conference on Dengue and Dengue Haemorrhagic Fever*, Pang, T. and Rathnanathan, R., Eds., Kuala Lumpur, Malaysia, 1983, 437.
20. Ussery, M. A., Burke, D. S., Nisalak, A., Andre, R. G., Leake, C. J., Elwell, M. R., and Tingpak-pongse, M., Isolation of Japanese encephalitis strains from patients, pigs and mosquitoes in Kampangphet Province, Thailand, during the 1982 Epidemic Season, *Proc. 32nd Ann. Meet. Am. Soc. Trop. Med. Hyg.*, San Antonio, Texas, 4 to 8 December, 1983.
21. Leake, C. J., Nisalak, A., Hoke, C. H., and Burke, D. S., Unpublished data, 1983, 1984.
22. Johnson, R., and Leake, C. J., Unpublished data, 1984.
23. Igarashi, A., Isolation of a Singh's *Aedes albopictus* cell clone sensitive to dengue and chikungunya viruses, *J. Gen. Virol.*, 40, 531, 1978.
24. Kuno, G., Dengue virus replication in a polyploid mosquito cell culture grown in serum-free medium, *J. Clin. Microbiol.*, 16, 851, 1982.
25. Leake, C. J., Unpublished data, 1983, 1984.

Chapter 14

ARBOVIRUS REPLICATION IN CELL CULTURES OF NONHEMATOPHAGOUS MOSQUITOES

G. Kuno

TABLE OF CONTENTS

I.   Introduction 44

II.   Isolation of Cell Lines 44

III.   Characteristics of Cell Lines 44
     A.   Morphology 44
     B.   Karyology 44
     C.   Isozyme 44
     D.   Adaptation to Serum-Free Media 45

IV.   Arbovirus Replication 45
     A.   Togaviridae *Alphavirus* 46
     B.   Flaviviridae *Flavivirus* 46
     C.   Rhabdoviridae *Vesiculovirus* 47
     D.   Bunyaviridae *Bunyavirus* and *Phlebovirus* 47
     E.   Reoviridae *Orbivirus* 48
     F.   Cytopathic Effects (CPE) 48
     G.   Plaque Development 48
     H.   Persistent Infection 48
     I.   Hydroxy and Carboxylic Acid Metabolism 48

V.   Applications to Biomedical Research 49
     A.   Virologic Surveillance for Dengue 49

References 52

I. INTRODUCTION

It is logical to employ cell lines derived from natural vectors for studies on arbovirus replication in vitro. The idea that cell lines derived from nonhematophagous mosquitoes can also be used for arboviral studies was conceived when it was found that dengue viruses could be propagated artificially by intrathoracic inoculation of the viruses in *Toxorhynchites amboinensis*, a predatory, nonhematophagous mosquito.[22] The first cell lines from the nonhematophagous mosquito reported in 1980 by two groups (Yale University and the San Juan Laboratories of Centers for Disease Control) provided opportunities to examine the utility of and application of these lines for arbovirus research.

II. ISOLATION OF CELL LINES

As with the isolation of most mosquito cell lines, eggs or larvae were found to be the ideal source of cells for initiating cell lines from nonhematophagous mosquitoes. Thus, Tesh[1] was able to isolate the TA-42 cell line from crushed eggs and the TA-9 cell line from fragments of first instar larvae of *T. amboinensis*. I found physically damaged, but live, whole, first instar larvae of *T. amboinensis* (Figure 1) a far more reliable source of seed cells from which to initiate cell lines and thus I readily obtained 22 continuous lines.[2,3] The cell line isolation rate was proportional to the number of physically damaged larvae per culture (Table 1). Although all of the above cell lines form monolayers, a line isolated by Munderloh et al.[4] from the embryos of *T. amboinensis* (RUTAE 12 V) has been grown as suspended multicellular vesicles of various sizes. All of the above *Toxorhynchites* cell lines were grown either in MM/VP 12 medium[5] or in L-15 medium[6] supplemented with tryptose phosphate broth, antibiotics, and fetal bovine serum (FBS). The relative ease with which many cell lines have been isolated from *T. amboinensis* may be because its eggs are larger than those of other mosquito genera.[1] In addition, growth-promoting molecules may be released from the physically damaged larvae.[3]

III. CHARACTERISTICS OF CELL LINES

A. Morphology

Most of the cell lines derived from *T. amboinensis* that form monolayers are composed primarily of fibroblast-like cells with a smaller number of epithelioid-type cells.

B. Karyology

Like those from other mosquitoes, cell lines from *T. amboinensis* such as TRA-171 and TA-42, contain predominantly diploid (2N = 6) cells. On the other hand, other cell lines such as TRA-272 exhibit more karyologic heterogeneity.[3] Repeated passages and cloning often lead to selection of particular kinds of cells. Thus, clones derived from the TRA-171 cell line were comprised predominantly of cells with 18 to 21 chromosomes.[7]

C. Isozyme

The isozyme profile is an important source of information for characterization of insect cell lines. Brown and Knudson[8] compared four isozymes (isocitrate dehydrogenase, malic enzyme, phosphoglucoisomerase, and phosphoglucomutase) in tick and mosquito cell lines and concluded that the TA-9 and TA-42 cell lines were identical with respect to isozyme mobility and that both lines were clearly different from other mosquito and tick cell lines tested.

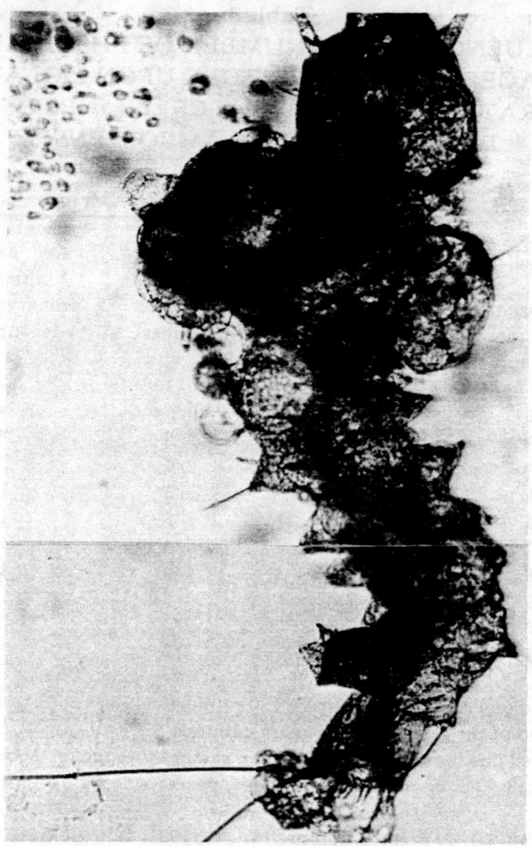

FIGURE 1.   A physically damaged but live larva of *Toxorhynchites amboinensis* in primary culture. (From Kuno, G., *In Vitro*, 16, 916, 1980. With permission.)

D. Adaptation to Serum-Free Media

The TRA-284 cell line was adapted to a serum-free medium composed of equal volumes of L-15 and 3% (W/V) tryptose phosphate broth (TPB).[9] Further study showed that most cell lines of both hematophagous and nonhematophagous mosquitoes could be adapted to such simple serum-free media in which equal volumes of TPB and of L-15, Eagle's minimal essential medium, or Medium 199 with Hanks' salts were mixed.[10]

The adaptation of cell lines to serum-free medium generally resulted in slower growth of cells, but lower cost of cell culture. Since the quality of bovine sera commonly used as a supplement in animal cell cultures often varies from lot to lot even from a single source of supply,[11] adaptation of cell lines to serum-free media meant the elimination of such exogenous factors as complement, antibodies to some arboviruses, and virus or most *Mycoplasma* contaminants, all of which could adversely affect studies of arboviruses in cell culture.

IV. ARBOVIRUS REPLICATION

Experimental data indicate that the cell lines of *T. amboinensis* are susceptible to infection with viruses naturally transmitted not only by mosquitoes but also by sandflies. Since many of the in-depth studies with these cell lines have been done only with the dengue viruses and related flaviviruses, further studies are needed to document susceptibility of the cells to arboviruses of other groups.

Table 1

INFLUENCE OF THE NUMBER OF PHYSICALLY
DAMAGED LARVAE PER TUBE UPON ISOLATION
RATE OF CONTINUOUS CELL LINES OF
TOXORHYNCHITES AMBOINENSIS

		Cell lines	
No. incapacitated larvae/tube	Avg. no. cell colonies/tube[a]	Isolation rate (no. isolates/ no. tubes)	Avg. days from initiation of primary culture to 1st subculture
1	0.7	0/3	—
2	1.0	0/5	—
3	2.6	2/5	172
4	2.7	2/6	232
5	3.3	3/3[c]	125
6	6.8	3/5	123
7	6.8	2/4[b]	151
8	8.5	1/4	104
9	7.6	4/5	95
10	17.3	3/3	79
11	14.5	2/2	81

[a]   One colony refers to a group of 150 or more cells attached to the glass at the end of the 4th week in primary culture.

[b]   One cell line discontinued after three passages because of poor growth.

[c]   Two cells lines discontinued after three passages because of poor growth.

From Kuno, G., *J. Med. Entomol.*, 18, 141, 1981. With permission.

A. Togaviridae *Alphavirus*

Tesh[1] infected both TA-42 and TA-9 cell lines with less than 2 $\log_{10}$ plaque forming units PFU/mℓ (as titrated in Vero cell cultures) of five alphaviruses and over 12 days of incubation observed that titers of Eastern equine encephalitis (EEE), Ross River, and Getah increased from 5 to 7 $\log_{10}$ PFU/mℓ in the TA-42 cell line, while remaining the same in the TA-9 cell line. Titers of Sindbis and chikungunya (CHIK) viruses, on the other hand, declined in both cell lines after 12 days of incubation. The results suggested that the TA-42 and TA-9 cell lines were not as sensitive as hematophagous mosquito cell lines that generally yield *Alphavirus* titers greater than 8 $\log_{10}$ PFU/mℓ. However, Legrand and Hotta[7] observed replication of the CHIK virus to over 10^9 PFU/mℓ after a 3-day incubation in a clone of the TRA-171 cell line.

B. Flaviviridae *Flavivirus*

Dengue viruses have been most extensively studied in nonhematophagous mosquito cell lines. Using the TRA-171 cell line, I demonstrated that tissue culture-adapted DEN-1 (Hawaii), DEN-2 (New Guinea "C"), and DEN-4 (H-241) replicated to titers of 6.0 $\log_{10}$ PFU/mℓ or above in 7 days.[12] These titers were lower, however, than the corresponding titers in simultaneously infected C6/36 (*Aedes albopictus*)[13] or AP-61 (*A. pseudoscutellaris*)[14] cell cultures (Table 2). The rate of increase in virus titer was also slower in the TRA-171 cell line. Suckling mouse-adapted DEN-3 (PR-6) did not replicate as well in any of the three cell lines and yielded titers of about 5 $\log_{10}$ PFU/ mℓ (Table 2). Legrand and Hotta[7] cloned the TRA-171 cell line and obtained DEN-2 (New Guinea "C") and DEN-4 titers of about 9 $\log_{10}$ PFU/mℓ and a DEN-3 (H-87) titer of 6 $\log_{10}$ PFU/mℓ. On the other hand neither TA-42 nor TA-9 cell lines supported

Table 2
REPLICATION OF DENGUE VIRUSES IN THREE MOSQUITO CELL LINES, INCLUDING THE TRA-171 CELL LINE FROM *T. AMBOINENSIS*

Virus	Multiplicity of infection (PFU/cell)	Cell line	Extracellular virus titer (days after inoculation)						
			0	1	2	3	4	5	7
			Log$_{10}$ PFU/ml supernatant fluid						
DEN-1 (Hawaii)	0.0006	AAL[a]	—[b]	2.18	3.18	4.90	5.18	5.40	7.53
		APS[c]	—	—	3.08	1.70	2.90	4.82	6.90
		TRA[d]	—	—	2.60	4.00	5.92	4.30	6.37
DEN-2 (NG "C")	0.0002	AAL	—	—	2.48	4.00	4.70	5.20	7.15
		APS	—	—	2.18	2.18	2.65	3.70	6.45
		TRA	—	—	—	3.91	4.40	3.78	6.00
DEN-3 (PR-6)	0.001	AAL	—	—	2.00	—	2.00	3.88	5.00
		APS	—	—	—	—	—	3.43	5.70
		TRA	—	—	—	—	2.43	5.13	4.70
DEN-4 (H-241)	0.008	AAL	—	3.04	3.30	4.54	5.30	6.40	7.58
		APS	—	—	2.70	—	3.11	4.70	7.62
		TRA	—	—	—	2.70	2.00	5.15	6.28

[a] AAL: *Aedes albopictus* cells (clone C6/36) by Igarashi.[13]
[b] No detectable titer.
[c] APS: *Aedes pseudoscutellaris* cells by Varma et al.[14]
[d] TRA: *Toxorhynchites amboinensis* cells (TRA-171) by Kuno.[2]

From Kuno, G., *In Vitro*, 17, 1012, 1981. With permission.

dengue virus replication.[1] In the latter study, the Kunjin virus also failed to replicate in the TA-42 cell line, while titers of Japanese encephalitis and West Nile viruses increased by over 1000-fold to 5.2 to 5.9 log$_{10}$ PFU/ml, respectively. In another study, the St. Louis encephalitis virus replicated in the TRA-171 cell line to titers exceeding 6 log$_{10}$ PFU/ml;[12] the titers in the TA-42 and TA-9 cell lines reached only 5.4 log$_{10}$ PFU/ml. The yellow fever virus (17D) replicated in the TRA-171 cell line to a titer of 4.48 log$_{10}$ PFU/ml.[12]

C. Rhabdoviridae *Vesiculovirus*

The vesicular stomatitis virus (New Jersey) replicated well in both the TA-42 and TRA-171 cell lines to titers of 7.5 and 6.48 log$_{10}$ PFU/ml, respectively.[1,12] Chandipura virus titers in the TA-42 and TA-9 cell lines also increased by 100- to 1000-fold to 5.6 and 4.2 log$_{10}$ PFU/ml.[1]

D. Bunyaviridae *Bunyavirus* and *Phlebovirus*

Titers of La Crosse, San Angelo, and Cache Valley bunyaviruses either remained the same or increased only slightly in the TA-9 cell line in 12 days; none of those viruses replicated in the TA-9 or TA-42 cell lines in 12 days.[1]

None of the five *Phleboviruses* tested by Tesh (Karimabad, Punto Toro, Pacui, sandfly fever [Sicilian], and Arumowot) replicated significantly in TA-42 or TA-9 cultures; titers of only the Karimabad virus remained about the same in both TA-42 and TA-9 cell lines, whereas titers of the other viruses declined sharply.[1]

E. Reoviridae *Orbivirus*

Changuinola and Corriparta viruses replicated in both the TA-42 and TA-9 cell lines and produced high virus titers of 7.5 $\log_{10}$ PFU/mℓ or above.[1]

F. Cytopathic Effects (CPE)

The most common CPE caused by arboviruses in *Toxorhynchites* cell lines was syncytial development by dengue viruses in the TRA-171 cell line.[2] A subline cloned from these cells was found to produce more consistent CPE.[7] In one experiment, ten human sera induced syncytia in TRA-284-SF (a subline of the TRA-284 cell line adapted to a serum-free medium) cell cultures and dengue viruses were recovered from all of them. The virus dose apparently influenced syncytial development, since nine of these sera contained 14 PFU or more per inoculum. In contrast, only one serum containing no demonstrable PFU per inoculum (as assayed by rhesus monkey kidney cell culture LLC-MK$_2$) induced syncytia.[9]

G. Plaque Development

Dengue viruses were found to cause plaques in the monolayer cultures of the TRA-171 cells.[12] The first overlay medium consisted of 0.75% agarose (Seaplaque brand, Marine Colloids, Inc., Rockland, ME),* 0.25% purified agar (Difco, Detroit, MI), 10% FBS, 1.5% essential amino acids (100 ×), 1.5% vitamin mixture (100 ×), 1.4% glutamine (200 mM), and 0.1% sodium bicarbonate (7.5%) in L-15 medium without phenol red. After 7 days of incubation with the first overlay medium at 32°C, the flasks were overlaid with the second medium containing 7 mℓ of neutral red (1:300) per 100 mℓ of the first plaquing medium. Plaques were counted after the flasks had been incubated at 35°C for 1 to 2 days. Plaque titers obtained in the monolayer cultures of the TRA-171 cells were approximately ten times lower than those obtained with LLC-MK$_2$ cells.[12]

H. Persistent Infection

Dengue viruses can cause persistent infections (PI) in the TRA-171 cell line (Figure 2).[15] Most often viruses continuously induced CPE, in particular syncytia and large globular cells. After a year of PI, antigenic characteristics of the viruses differed from those inoculated as revealed by plaque-reduction neutralization test (Table 3). The viruses in the PI cultures were also temperature-sensitive. Further changes in the properties of dengue viruses in the PI cultures were manifested as reduction in neurovirulence in suckling mice. It has been reported that Sindbis virus-infected *Aedes albopictus* cell cultures release molecules into the supernatant fluid that interfere with the replication of related viruses.[16] However, virus-free supernatant fluids of TRA-171 cell cultures persistently infected with dengue viruses did not interfere with the replication of other dengue serotypes.[15]

I. Hydroxy and Carboxylic Acid Metabolism

Studies of hydroxy acids in the supernatant fluids of the dengue virus infected TRA-284-SF cell cultures by means of frequency-pulsed electron capture gas-liquid chromatography revealed that DEN-1 and DEN-4 induced serotype-specific profiles.[17] DEN-2 and DEN-3 profiles of hydroxy acids were indistinguishable from each other,

---

* Mention of trade names and commercial sources is for identification only and does not constitute the endorsement of the U.S. Department of Health and Human Services.

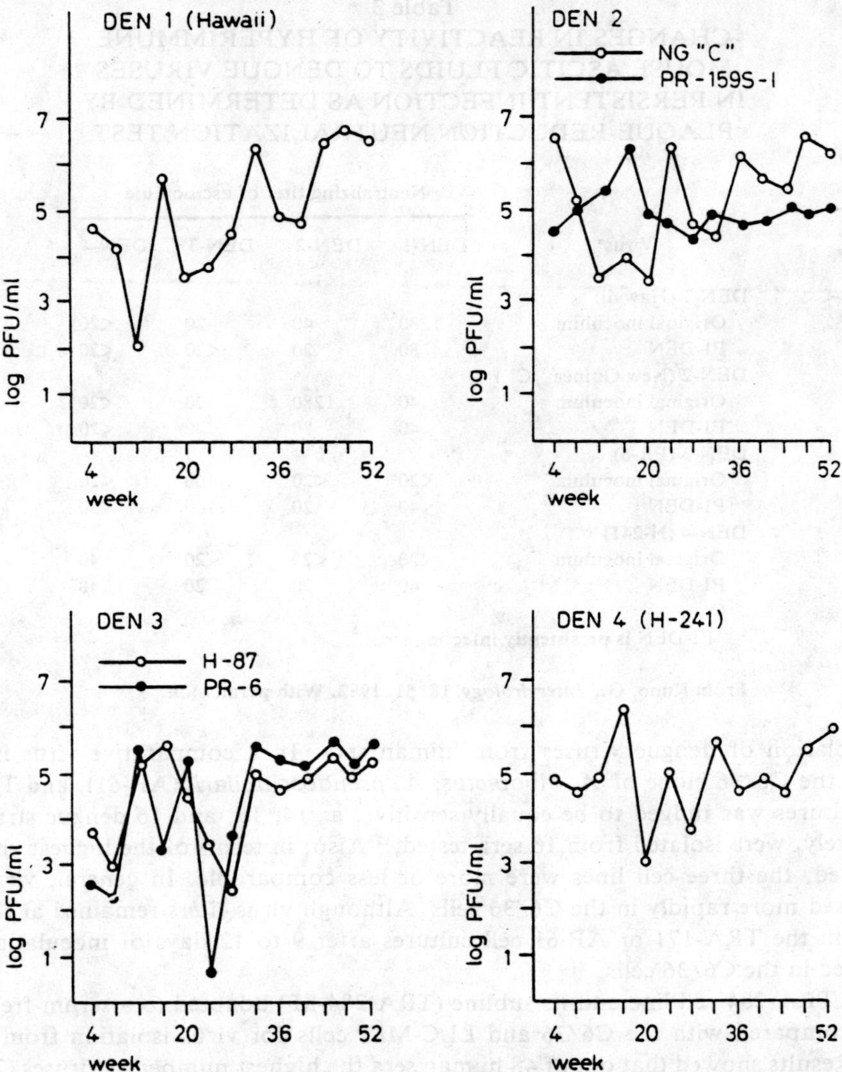

FIGURE 2.   Fluctuation of extracellular virus titer in the persistently infected TRA-171 cell cultures infected with dengue viruses. (From Kuno, G., *Intervirology*, 18, 47, 1982. With permission.)

but differed from those of uninfected controls and from DEN-1 or DEN-4 profiles. Carboxylic acid profiles in cultures infected with four serotypes were different from that of the uninfected control culture, but none of them were serotype specific.[17] These results contrast with previous findings in dengue-infected LLC-MK₂ cells that all serotypes induced serotype-specific hydroxy acid profiles and a DEN-1-specific carboxylic acid profile.[18]

V. APPLICATIONS TO BIOMEDICAL RESEARCH

A. Virologic Surveillance for Dengue

The most important application of nonvector mosquito cell cultures has been for use in isolating viruses from clinical specimens. So far, only the C6/36 clone of Singh's *Aedes albopictus* cells, AP-61, TRA-171, and TRA-284-SF cell lines has been used for

Table 3
CHANGES IN REACTIVITY OF HYPERIMMUNE
MOUSE ASCITIC FLUIDS TO DENGUE VIRUSES
IN PERSISTENT INFECTION AS DETERMINED BY
PLAQUE-REDUCTION NEUTRALIZATION TEST

	Neutralizing titer of ascitic fluid			
Virus[a]	DEN-1	DEN-2	DEN-3	DEN-4
DEN-1 (Hawaii)				
Original inoculum	1280	40	20	<20
PI-DEN	80	20	<20	<20
DEN-2 (New Guinea "C")				
Original inoculum	40	1280	20	<20
PI-DEN	40	80	<20	<20
DEN-3 (PR-6)				
Original inoculum	<20	<20	160	<20
PI-DEN	40	20	160	<20
DEN-4 (H-241)				
Original inoculum	<20	<20	<20	40
PI-DEN	40	20	20	40

[a]   PI-DEN is persistently infecting virus.

From Kuno, G., *Intervirology,* 18, 51, 1982. With permission.

the isolation of dengue viruses from human sera. In a comparative virus isolation study, the C6/36 clone of *A. albopictus, A. pseudoscutellaris* (AP-61), and TRA-171 cell cultures was judged to be equally sensitive, as 14, 15, and 16 dengue strains, respectively, were isolated from 16 sera tested.[12] Also, in terms of the highest virus titer obtained, the three cell lines were more or less comparable. In general, virus titers increased more rapidly in the C6/36 cells. Although virus titers remained at the same levels in the TRA-171 or AP-61 cell cultures after 9 to 12 days of inoculation, they dropped in the C6/36 cells.

The TRA-284 cell line and its subline (TRA-284-SF) adapted to a serum-free media were compared with the C6/36 and LLC-MK$_2$ cells for virus isolation from human sera. Results showed that out of 48 human sera the highest number of viruses (25) were isolated in the TRA-284 cells, followed by 22 in the TRA-284-SF cells, 17 in the C6/36 cells, and 10 in the LLC-MK$_2$ cells.[9] These results suggested that sensitivity to dengue virus infection was not lost through cellular adaptation to a serum-free medium. Since such media are generally more economical and readily available than those containing serum, a cell line adapted to serum-free media is quite useful for diagnosis of viral disease in areas where good quality bovine serum is either too expensive or difficult to acquire. When rates for isolation of dengue viruses from human sera were compared among the TRA-284-SF, C6/36, and AP-61 cell lines, the highest (36/55) was observed for the first line (Table 4).[19] The virus isolation rate was also the function of the amount of virus per inoculum (Table 5) since greater numbers of inocula containing less than 1 PFU/0.1 mℓ inoculum failed to infect the TRA-284-SF cell cultures (Table 5).[9] In a similar comparative study of dengue virus isolation conducted in Thailand by Leake et al., the TRA-284-SF cell cultures were the most sensitive among three cell lines tested.[20]

Unpassaged DEN-1 and DEN-2 viruses were often detected earlier in the C6/36 cells than in the other cell lines, but viral infection did not noticeably spread to other cells in the C6/36 cultures.[21] With unadapted DEN-3 and DEN-4 strains, virus growth in

Table 4
COMPARATIVE DENGUE VIRUS ISOLATION FROM CLINICAL SPECIMENS (HUMAN SERA) IN 3 MOSQUITO CELL LINES[a]

Cell Line	Experiment 1		Experiment 2		Total[b]	
	Number	%	Number	%	Number	%
C6/36	29/100[c]	29	14/55	25	43/155	28
AP-61	34/100	34	17/55	31	51/155	33
TRA-284	33/100	33	—	—	33/100	33
TRA-284-SF	—	—	20/55	36	20/55	36

[a] All cells grown in tube cultures except the TRA-284, which was grown in flasks.

[b] All isolates were dengue 1 and dengue 4.

[c] Number isolates/number tested.

Table 5
RELATIONSHIP BETWEEN THE AMOUNT OF DENGUE VIRUS IN HUMAN SERUM AND VIRUS ISOLATION IN *T. AMBOINENSIS* AND LLC-MK$_2$ CELL CULTURES

Dengue serotype	No. of sera tested	Amount of virus per inoculum[a]	No. of virus strains isolated in following cells	
			TRA-284-SF	LLC-MK$_2$
DEN-1	17	1	12	4
	8	1—100	7	3
	3	>100	3	3
DEN-2	4	1—100	4	4
	2	>100	2	2
DEN-3	2	<1	2	0
	4	1—100	2	1
DEN-4	4	<1	2	0
	4	1—100	2	3

[a] PFU per 0.1 mℓ of undiluted serum per flask.

From Kuno, G., *J. Clin. Microbiol.*, 16, 854, 1982. With permission.

the C6/36 cells was inferior to that in the TRA-284-SF or AP-61 cells, particularly when virus titers of the inocula were <2.0 tissue culture infectious dose$_{50}$ per mℓ.[20] Thus, the TRA-284-SF cell cultures provide not only high efficiency of dengue virus isolation, but an economic means of viral diagnosis.

As a practical application to virologic surveillance for dengue in Puerto Rico, the TRA-284-SF cells are grown as three million cells/3 mℓ of medium in disposable, screw cap, glass tubes (16 × 125 mm) in a slanted rack. When monolayers are confluent, they are inoculated with 0.05 mℓ aliquots of undiluted human serum. Cultures are incubated for 7 to 10 days and the cells are harvested and tested for virus infection by immunofluorescence. Viruses are identified using monoclonal antibodies (see Chapter 11, this volume). Such a virologic surveillance was found to be economical and advantageous for processing a large number of clinical specimens in a short time.[21]

REFERENCES

1. Tesh, R. B., Establishment of two cell lines from the mosquito *Toxorhynchites amboinensis* (Diptera: Culicidae) and their susceptibility to infection with arboviruses, *J. Med. Entomol.*, 17, 338, 1980.
2. Kuno, G., A continuous cell line of a nonhematophagous mosquito, *Toxorhynchites amboinensis*, *In Vitro*, 16, 915, 1980.
3. Kuno, G., A method for isolating continuous cell lines from *Toxorhynchites amboinensis* (Diptera: Culicidae), *J. Med. Entomol.*, 18, 140, 1981.
4. Munderloh, U. G., Kurtti, T. J., and Maramorosch, K., *Anopheles stephensi* and *Toxorhynchites amboinensis*: aseptic rearing of mosquito larvae on cultured cells, *J. Parasitol.*, 68, 1085, 1982.
5. Varma, M. G. R. and Pudney, M., The growth and serial passage of cell lines from *Aedes aegypti* (L.) larvae in different media, *J. Med. Entomol.*, 6, 432, 1969.
6. Leibovitz, A., The growth and maintenance of tissue cultures in free gas exchange with the atmosphere, *Am. J. Hyg.*, 78, 173, 1963.
7. Sanchez Legrand, F. and Hotta, S., Susceptibility of cloned *Toxorhynchites amboinensis* cells to dengue and chikungunya viruses, *Microbiol. Immunol.*, 27, 101, 1983.
8. Brown, S. E. and Knudson, D. L., Characterization of invertebrate cell lines. IV. Isozyme analyses of Dipteran and acarine cell lines, *In Vitro*, 18, 347, 1982.
9. Kuno, G., Dengue virus replication in a polyploid mosquito cell culture grown in serum-free medium, *J. Clin. Microbiol.*, 16, 851, 1982.
10. Kuno, G., Cultivation of mosquito cell lines in serum-free media and their effects on dengue virus replication, *In Vitro*, 19, 707, 1983.
11. Price, P. J. and Gregory, E. A., Relationship between in vitro growth promotion and biophysical and biochemical properties of the serum supplement, *In Vitro*, 18, 576, 1982.
12. Kuno, G., Replication of dengue, yellow fever, St. Louis encephalitis and vesicular stomatitis viruses in a cell line (TRA-171) derived from *Toxorhynchites amboinensis*, *In Vitro*, 17, 1011, 1981.
13. Igarashi, A., Isolation of Singh's *Aedes albopictus* cell clone sensitive to dengue and chikungunya viruses, *J. Gen. Virol.*, 40, 531, 1978.
14. Varma, M. G. R., Pudney, M., and Leake, C. J., Cell lines from larvae of *Aedes (Stegonyia) malayensis* Colless and *Aedes* (S.) *pseudoscutellaris* (Theobald) and their infection with some arboviruses, *Trans. R. Soc. Trop. Med. Hyg.*, 68, 374, 1974.
15. Kuno, G., Persistent infection of a nonvector mosquito cell line (TRA-171) with dengue viruses, *Intervirology*, 18, 45, 1982.
16. Riedel, B. and Brown, D. T., Novel antiviral activity found in the media of Sindbis virus-persistently infected mosquito *(Aedes albopictus)* cell cultures, *J. Virol.*, 29, 51, 1979.
17. Kuno, G., Brooks, J. B., and Wycoff, B. J., Changes in hydroxy and carboxylic acid compositions in the supernatant fluids of mosquito cell cultures infected with dengue viruses, *J. Invertebr. Pathol.*, 44, 256, 1984.
18. Brooks, J. B., Kuno, G., Craven, R. B., and Alley, C. C., Studies of metabolic changes in cell cultures infected with four serotypes of dengue fever viruses by frequency-pulsed electron capture gas-liquid chromatography, *J. Chromatogr.*, 276, 279, 1983.
19. Kuno, G., Gubler, D. J., Velez, M., and Oliver, A., Comparative sensitivity of three mosquito cell lines for isolation of dengue viruses, *Bull. W.H.O.*, 63, 279, 1985.
20. Leake, C. J., Nisalak, A., and Burke, D. S., Comparative isolation of dengue viruses from DHF patients by mosquito inoculation and on three mosquito cell lines, in *Proceeding of International Conference on Dengue Fever and Dengue Haemorrhagic Fever*, Pang, T. and Rathnanathan, R., Eds., Kuala Lumpur, Malaysia, 1983, 437.
21. Gubler, D. J., Kuno, G., Sather, G. E., Velez, M., and Oliver, A., Surveillance for dengue viruses using mosquito cell cultures and specific monoclonal antibodies, *Am. J. Trop. Med. Hyg.*, 33, 158, 1984.
22. Gubler, D. J., Personal communication.

Chapter 15

A COMPARATIVE VIEW OF THE REPLICATION OF TOGAVIRUSES IN VERTEBRATE AND MOSQUITO CELLS

P.-J. Enzmann

TABLE OF CONTENTS

I.   Introduction 54

II.   Molecular Biology of the Viruses 54
      A.   Processing and Modification of Viral mRNAs 54
      B.   Physical Properties of the *Togavirus* Genome 54
      C.   Viral Structural Proteins 55

III.   Organization of the Virion 55

IV.   Growth Cycle of Togaviruses 56

V.   Early Events in Virus Replication 56

VI.   Viral Protein Synthesis 57

VII.   Viral RNA Synthesis 58

VIII.   Assembly of the Nucleocapsid 59

IX.   Formation of the Viral Envelope 59

X.   Budding Process 60

XI.   Similarities and Differences in Vertebrate and Arthropod Cells 60

XII.   Effects of Virus Replication on Host Cell Macromolecular
       Synthesis 62

References 63

I. INTRODUCTION

The present review concentrates on the replication of togaviruses in vertebrate and mosquito cells mainly as a comparison of the effects which occur during and after infection of these different host cells. The great number of publications on the multiplication of togaviruses precludes a complete coverage of existing literature within the frame of this article, hence this review will be selective. Review articles are cited whenever possible.

In 1964, Mussgay[1] stated: "According to the definition one has to distinguish between two main host systems of arboviruses, namely, cells of vertebrates and cells of arthropods. It would be interesting to compare the mode of arbovirus multiplication in these phylogenetically different systems at the cellular level. Unfortunately, however, this cannot be done because there are no known methods for preparing arthropod tissue cultures which allow quantitative work. Therefore, an attempt to summarize knowledge of the growth cycle of arboviruses in vertebrate and arthropod cells is by necessity concerned with an in vitro system, i.e., tissue culture of vertebrate cells, and in vivo systems, i.e., vertebrate hosts and whole arthropods." In the meantime, significant advances have been made first by Singh[2] in the development of arthropod cell lines, mainly those derived from the larvae of *Aedes albopictus* and *A. aegypti*, and then by others in increasing numbers and varieties during the last 15 years. The history and methods of invertebrate tissue culture have been reviewed by many,[3-7] and an overview of this topic is given in another chapter.

Cell cultures derived from known virus vectors, namely members of the phyla Diptera and Acari, are of main interest in this respect. In the present review only established cell lines are considered because they are the most suitable system for studying virus replication; for studies dealing with arboviruses in surviving invertebrate tissue fragments and in primary cell cultures, the reader is referred to a review by Yunker[8] and to other chapters of this book.

II. MOLECULAR BIOLOGY OF THE VIRUSES

Togaviruses belong to the group of positive strand RNA viruses (plus-stranded). This means that the virion RNA is able to function as a messenger RNA and as a consequence, virion RNA is infectious. Initiation of infection is performed by translation of the parental genomic RNA to produce the viral replicase and transcriptase enzymes. By this enzyme complex, minus-strand templates are transcribed and plus-strand genomes are replicated; in addition, in the case of alphaviruses, plus-strand subgenomic mRNAs for virion structural proteins are generated.[9]

A. Processing and Modification of Viral mRNAs

Alphavirus mRNAs are polyadenylated (about 50 to 200 nucleotides long) at their 3′-termini. The 3′-polyadenylation in mRNA can arise by transcription of 5′-terminal polyuridylate sequences in complementary negative strands or by posttranscriptional additions. *Alphavirus* mRNAs are capped at the 5′-termini principally with the methylated cap 1 structure $m^7G(5')ppp(5')N^mNp$.[10,11]

B. Physical Properties of the *Togavirus* Genome

Polyacrylamide gel electrophoresis of the *Alphavirus* genome gave estimates of 4.0 to 4.4×10^6 and sedimentation analysis resulted in 4.4×10^6 corresponding to 40 to 49S.[12-16] Most accurate sedimentation analysis of the Sindbis virus RNA was performed by Simmons and Strauss[13] who obtained 49S. However, as a convenience, the *Alphavirus* genome will be designated 42S. The RNA of flaviviruses has been less carefully characterized than that of alphaviruses. It has a sedimentation coefficient of approximately 45S.[16] The *Flavivirus* RNA is capped, but lacks poly (A).[17]

C. Viral Structural Proteins

In the genus *Alphavirus,* virions are composed of two glycoproteins in the viral envelope (E1 and E2) and one capsid protein (C) within the range of 50 kdalton (E1 and E2) and 30 kdalton (C), respectively. An additional glycoprotein, (E3), with a molecular weight of about 10 kdalton, has been found in Semliki Forest virion.

Flaviviruses are composed of one large envelope glycoprotein named V3 (50 to 60 kdalton), one small, nonglycosylated, membrane-associated protein V1 (7 to 8 kdalton), and a nucleocapsid protein, V2 (13 kdalton).[16,18] From nucleotide sequencing, the amino acid sequences of the structural proteins of Sindbis and Semliki Forest viruses were determined. The E1 protein of the Semliki Forest virus consists of 438 amino acids and that of the Sindbis virus consists of 439 residues. There is one complex oligosaccharide chain attached to the E1 protein of Semliki Forest virus and two sugar units to that of Sindbis virus.

The E2 protein of the Semliki Forest virus consists of 422 amino acids and that of the Sindbis virus consists of 423 amino acids and in each case carries two sugar units. The E3 protein of the Semliki Forest virus consists of 66 amino acids and carries one sugar unit.[19]

III. ORGANIZATION OF THE VIRION

The togaviruses are the smallest enveloped animal viruses; alphaviruses are about 60 nm in diameter; flaviviruses are about 45 nm when analyzed by electron microscopy. Papers reviewing the structure of togaviruses have been published recently.[18-23] This size difference is also reflected by the sedimentation coefficients which range between 240 and 300S for alphaviruses and 170 to 220S for flaviviruses. The buoyant density depends on the gradient material used and varies between 1.18 (in sucrose) and 1.24 g/ mℓ (in CsCl) for viruses of both genera. In general, the togaviruses consist of an isometric internal structure, termed nucleocapsid or core, and an envelope with the electron microscopic characteristics of a membrane. The envelope may display a smooth surface or may carry more or less prominent projections.

In detail, symmetrically arranged subunits have been demonstrated on *Alphavirus* nucleocapsids which appear to be composed of 32 capsomers arranged in a T=3 icosahedral surface lattice.[24] The glycoproteins of the Sindbis virus are also arranged in an icosahedral symmetry in the viral envelope.[25] It could be demonstrated that the 80 subunits of the envelope display a hexon-penton clustering in a T=4 icosahedral surface lattice. The glycoprotein projections are located in bundles on the surface of the morphological subunits. In spite of these differences in the symmetrical arrangement of subunits in nucleocapsid and envelope, a direct contact between each nucleocapsid protein and the corresponding envelope proteins is possible and is consistent with an equimolar ratio of structural proteins.[24] This evidence was postulated by the fact that the viral glycoproteins have hydrophobic tails and penetrate the lipid bilayer for interaction with the C-protein of the nucleocapsid.[19,26,27] It is worth mentioning that the nucleocapsid structure is influenced by the pH of the surrounding fluid. The diameter of the nucleocapsid was reduced from 39 to 33 nm after lowering the pH to 6.2.[28]

The morphology of flaviviruses is uniform and yet is distinct from that of the alphaviruses. The surface projections of flaviviruses appear to vary considerably in thickness and detail. The exact icosahedral arrangement of the subunits on the nucleocapsid is not clear. Only few structural details were described until now.[21]

IV. GROWTH CYCLE OF TOGAVIRUSES

Replication of togaviruses occurs over a wide range of temperatures between 15 and 41°C. Work stimulated by an interest in the growth of viruses of homeothermic vertebrates in cells of hosts far removed phylogenetically from the mammals, and at temperatures below those of mammals has demonstrated adaptation to and growth of togaviruses in mammalian, avian, reptilian, amphibian, fish, and arthropod cells.[16,29] With alphaviruses a one-step growth curve is completed at 37°C usually after 6 to 10 hr, whereas at lower temperatures ranging from 15 to 28°C, the time may be about twice as long (28°C) or in the extreme case (15°C) about 10 days.[30] Therefore, virus replication is related to the incubation temperature and to the optimum temperature of cell growth.

As with many in vitro cell systems, the efficiency of virus replication varies considerably, depending on the strains of virus and the cells used. With carefully selected virus strains and cell clones, one-step growth curves can be achieved with cultured mosquito cells which, apart from a slightly longer latent period, are similar to those observed with the same virus in vertebrate cells.[7]

Arthropod vectors, as a rule, show no pathological effect of infection, while naturally and experimentally infected vertebrates often do. These facts seem to be confirmed (especially for alphaviruses) in cell culture systems derived from arthropods and vertebrates. Infection is highly cytotoxic for permissive vertebrate cells, while permissive arthropod cells seldom display detectable cytopathic effects; normally the latter become persistently infected. This contrast between cytolytic infection of vertebrate cells and noncytolytic infection of arthropod cells is not absolute, since it is possible to establish chronically infected vertebrate cell cultures and conversely, some destructive processes may result from a virus infection also in arthropod cells (see Chapters 16 to 18, this volume).

Togavirus growth cycles are usually divided into three phases. During the first phase the virus is internalized by the host cell to be uncoated and capable of initiating replication. This first phase includes absorption of the viral particle by the cell membrane and penetration into the cell. The second phase begins with translation of the incoming viral genome followed by transcription and replication procedures and synthesis of the viral structural components. The third phase is characterized by assembly of the viral nucleocapsids from RNA and core-proteins and by specific replacement of cellular proteins in the host-cell plasma membrane by viral envelope proteins. Viral particles are liberated from the infected cells in a specific budding process.

V. EARLY EVENTS IN VIRUS REPLICATION

The first phase of *Togavirus* replication is started by adsorption of the virus to susceptible cells which occurs within a few minutes. Adsorption of alphaviruses is highly specific in its salt requirement in that maximal binding occurs only in the presence of defined concentrations of monovalent cations. The presence of divalent cations in the medium appears to inhibit virus attachment. This salt requirement for adsorption of alphaviruses may be a group characteristic because maximal attachment of flaviviruses onto cells occurs in the presence of divalent cations.[29,31] But adsorption is more than a nonspecific electrostatic interaction between negatively and positively charged groups on the virus and cell surfaces. This highly specific process is mediated through the recognition of receptors on the cell membrane by defined regions of the viral glycoproteins. The number of receptors for alphaviruses in chick and BHK cells has been determined to be about 10^5 per cell.[16] The nature of the receptor is still unknown. Different lipids have been suggested by several investigators.[16,23] It has also been reported that the specific receptor for Semliki Forest virus is the major histocompatibility antigen on the surface of mouse and human cells.[32] However, Oldstone et al.[33] demonstrated that this antigen is hardly involved in virus reception. The data on the mode of host cell entry for togaviruses are controversial. The findings of Helenius et al.[34] and Fan and Sefton[35] support the view that adsorptive endocytosis is the way to enter the cell. On the other hand fusion of the viral membrane with the cell plasmalemma[36] or pinocytosis[37] may be the initial event in virus penetration.

The second and third phases of *Togavirus* replication, i.e., protein- and RNA-synthesis, and morphogenesis of progeny virus particles are described below.

VI. VIRAL PROTEIN SYNTHESIS

When the alphaviral genome is uncoated about two thirds of it starting from the 5′-terminus is translated into one large protein. This is cleaved into four proteins with molecular weights of about 86 kdalton, 72 kdalton, 70 kdalton, and 60 kdalton, some or all of which form the *Alphavirus* RNA polymerase. This enzyme is necessary to transcribe the viral genome into minus-strands which act as templates for genome and 26S-RNA synthesis.

Protein produced from the nonstructural segment of the genome is also required for formation of the subgenomic 26S RNA. Both RNA-types act as mRNA for further protein synthesis. Translation of 42S RNA does not produce structural proteins. Thus, in common with other eucaryotic polycistronic mRNAs, the internal site for the initiation of synthesis of structural proteins is not functional in 42S RNA. Therefore, translation of 42S RNA terminates in the region of RNA prior to the structural protein RNA sequences. The 26S RNA represents the 3′-terminal one third of the genome RNA molecule and codes as a polycistronic messenger for the structural proteins which are translated from and in the form of a 130 kdalton precursor protein. The gene sequence of 5′-C-PE2-5kdalton-E1-3′ on the 26S RNA has been firmly established. C is the capsid protein, PE2 is a precursor for the envelope proteins E2 and E3 (E3 is not found in the virions of all viruses), and kdalton 5 is another peptide with a molecular weight between 4.5 and 6 kdalton (it does not appear in the virion).

The capsid protein C is the first protein synthesized by the 26S RNA. Soon after completion of its synthesis it is cleaved from the polyprotein (precursor). This cleavage seems to be catalyzed by the capsid protein itself.[38] C protein associates rapidly with the viral RNA to form nucleocapsids. The remaining structural proteins (3 in Semliki Forest virus) are two membrane glycoproteins which traverse the lipid bilayer by short stretches found at or near the C-terminus of the proteins.[39,40] The cleavage between the envelope proteins PE2 and E1 takes place at the time when most of the polyprotein has been translated. The first product is PE2, the precursor of E2, and then follows E1. This processing takes place after anchorage of the precursor in the endoplasmatic reticulum. For generation of PE2 and E1, two proteolytic cleavages on the large precursor are necessary. A short peptide with a molecular weight of 4.5 kdalton to 6 kdalton, depending on the virus strain, is cut out between PE2 and E1 with the function of this peptide not yet clear. Generation of E2 and E3 from PE2 seems to take place as a morphogenetic cleavage during budding of the viral particle. It depends on the virus strain whether E3 is incorporated as an integral part of the virion (Semliki Forest virus) or it is shed into the medium (Sindbis).

For effective cleavage of *Alphavirus* polyproteins, at least two proteases are necessary, a virus-coded activity present on the capsid protein which acts autocatalytically and one or possibly two cellular proteases. In an excellent review by Strauss and Strauss[9] the processes for generation of structural proteins are described.

The translation strategy of flaviviruses has not been definitely established. There is no evidence for a subgenomic RNA comparable with the 26S RNA of alphaviruses; it is generally believed that the viral RNA is the only messenger. Westaway[41] has proposed that the structural proteins V1 (membrane associated), V2 (capsid), and V3 (envelope), as well as the nonstructural proteins P20, P27, P37, P71, and P100, are separately initiated and terminated during translation, a process which is unusual in eukaryotes, but there are data supporting this hypothesis.[41] On the other hand there are also data which indicate that at least the structural proteins encoded in the 5′-end of the genome may be produced by posttranslational cleavage or trimming.[9]

VII. VIRAL RNA SYNTHESIS

After uncoating of the infecting *Alphavirus,* the incoming RNA serves as the initial messenger molecule to form nonstructural polypeptides, some or all of which are components of the virus-specified RNA-dependent RNA polymerase. Replication and transcription of viral RNA involves an initiation process followed by an elongation phase. It was postulated by Strauss and Strauss[9] that four activities are involved in replication and transcription of *Alphavirus* RNA: an elongation enzyme which synthesizes the RNA chain once initiated, and three initiation activities which initiate minus-strands, plus strands and the subgenomic mRNA. These activities could reside in the four different polypeptide chains which are synthesized in the first biochemical process. This hypothesis is supported by the fact that in the case of the Sindbis virus, four complementation groups are required for normal RNA synthesis after infection with temperature-sensitive mutants.[42] The genes for viral RNA replicase (nonstructural functions) are encoded near the 5′-end and require about two thirds of the viral RNA (42S RNA). The genes for the viral structural proteins reside in the 3′-third of the molecule. There is no evidence for early translation of the structural gene products.

After translation of the RNA replicase the incoming 42S RNA is transcribed by this enzyme as an early function in infection. Apparently the minus-strand replicase recognizes a structure in the genomic RNA and then starts minus-strand RNA synthesis.[43] This species of RNA is not found free in the infected cell but rather is tightly associated with the RNA-synthesizing apparatus where it functions as a template for plus-strand synthesis. This apparatus is a multistranded complex consisting of a 42S negative-strand RNA and several nascent plus-strands of varying length. During the whole replication cycle, only this type of negative-strand RNA appears. It contains a 5′-terminal tract of poly (U) complementary to the 3′-terminal poly (A) tract in 42S virion-RNA. The maximum rate of synthesis of negative-strand RNA is reached at about 150 min after infection, declining rapidly thereafter. The rate of synthesis of positive 42S RNA increases exponentially up to 3 hr after infection, remaining constant thereafter up to 6 to 7 hr. Regulation of minus-strand synthesis seems to be a viral function, but the nature of this process is still unclear. There is the possibility that plus-stranded RNA once synthesized is readily used as mRNA for protein synthesis or for nucleocapsid formation, therefore, only a very limited amount of plus-strands is used for transcription. Synthesis of plus-stranded 42S RNA is paralleled by synthesis of 26S RNA, a subgenomic mRNA which codes for the structural proteins. As mentioned above, the 26S RNA represents the 3′-terminal one third of the genome. This type of RNA is transcribed directly from 42S negative-stranded RNA, therefore, there must be a special replicase recognizing an internal initiation signal on the 42S RNA,[44] or the function of a special protein.[23] Nevertheless, regulation of 26S RNA is a viral function. It is an efficient mechanism of reiteration of a portion of the viral genome and allows excess synthesis of the viral structural proteins independently of the nonstructural proteins synthesis by the 42S RNA molecule.

Flavivirus RNA replication is not well understood. As with alphaviruses, the infecting RNA must be translated to produce the viral replicase, but the translation as well as the replication strategy of the *Flavivirus* genome has not been definitely established.

VIII. ASSEMBLY OF THE NUCLEOCAPSID

As repeatedly mentioned, alphaviruses and flaviviruses differ in many aspects during the replication cycle and must be regarded separately in the passage dealing with the assembly of nucleocapsid and virion. Again, there is more information about morphogenesis of alphaviruses compared with flaviviruses. Papers reviewing the assembly of *Alphavirus* nucleocapsid have been published recently.[9,19,21,22]

The capsid protein of alphaviruses is the first protein to be translated from the subgenomic 26S RNA which functions as a polycistronic mRNA for all structural proteins. Cleavage of the capsid protein from the nascent precursor appears to be a process catalyzed by the C-protein itself.[38] Newly synthesized C-protein associates immediately with 42S RNA to form nucleocapsids in the cytoplasm. Major interactions during nucleocapsid assembly seem to take place between protein and RNA, rather than between the protein molecules since neither empty nucleocapsids nor capsid protein aggregates without nucleic acid are formed during infection. *Alphavirus* RNA is therefore an integral part of the nucleocapsid structure, a fact which is also indicated by the sensitivity of the nucleocapsid towards treatments with RNase and low concentrations of SDS.[28]

How the actual packaging of viral RNA into the nucleocapsid proceeds is proposed by Strauss and Strauss:[9] there is a conserved sequence of 51 nucleotides within the genome of alphaviruses, located near the 5'-terminus of the RNA which forms a stable hairpin structure. One or both of these loops may be a binding site for C-protein which then acts as a nucleation site for encapsidation. If the C-protein is present, this process is favored above binding of the same structure to minus-strand replicase. Therefore, minus-strand synthesis would be regulated by nucleocapsid formation. On the other hand this theory may explain why 26S RNA is never encapsidated; it represents only the 3'-terminal one third of the genome and the double hairpin structure is situated near the 5'-terminus.

Nucleocapsids accumulate in the cytoplasm and migrate to the cytoplasmic side of the host cell membranes to get enveloped. Morphogenesis of flaviviruses was reviewed by Westaway[41] and Murphy.[21] There is no biochemical evidence for precursor particles comparable with the nucleocapsids of alphaviruses. There are several reports on the existence of precursor particles seen in electron microscopic pictures, but these have received little or no support. Usually mature virions were found in cytoplasmic vacuoles (for a review see Reference 41.)

IX. FORMATION OF THE VIRAL ENVELOPE

In the *Alphavirus* system next to the C-protein, a precursor of E2 protein called PE2 is translated from the 26S RNA, followed by E1. In contrast to the C-protein which is needed free in the cytoplasm for assembly of nucleocapsids, the envelope proteins have to be inserted into the membranes of the endoplasmatic reticulum for normal processing which includes glycosylation and transport. Removal of the capsid-protein from the nascent precursor polyprotein allows a signal sequence of about 19 residues at the N-terminus to function, and results in the integration of precursor PE2 in the endoplasmatic reticulum with concomitant glycosylation.[9] A second signal sequence with about 5 kdalton is situated between PE2 and E1. After completion of PE2 this signal sequence is integrated into the endoplasmatic reticulum giving the signal that a membrane protein (E1) is translated. Removal of this signal sequence which separates PE2 from E1 requires two proteolytic cleavages. Glycosylation of E1 occurs on the nascent protein-chain. The glycoproteins once synthesized and inserted into the endoplasmatic reticulum migrate to the plasma membrane by way of the Golgi apparatus. The cleavage of PE2 to form E2 and E3 has been postulated to occur in the Golgi vesicles.[45,46] There remains another way to process PE2. Cleavage of PE2 can be prevented by antiserum against E1[47,48] and antigenicity of PE2 is changed during maturation.[49,50] Cleavage of PE2 occurs only partially in infected mosquito cells treated with actinomycin D.[51] The only other circumstance in which cleavage of PE2 occurs in the absence of virus maturation is when infected cells are placed in media of low ionic strength.[52] Therefore processing of PE2 may be interpreted as a morphogenetic process including a reorganization of the glycoprotein which takes place after budding is initiated by the underlaying nucleocapsid.[22] In the *Flavivirus* system viral membrane proteins migrate to the membranes of the endoplasmatic reticulum and to a lesser extent to those of Golgi complex.[41]

X. BUDDING PROCESS

The final step in virus maturation is the coating of the nucleocapsid with the viral envelope at the plasma membrane of the infected host cell. The viral glycoproteins are synthesized and transported using the normal cellular mechanisms for these processes.[53] Virus capsids attach themselves to the inner surface of the cell membrane after modification of that membrane by insertion of the virus specific glycoproteins and exclusion of host proteins. This final stage in the virus assembly involves a specific interaction between the nucleocapsid and the modified cell membranes in that capsid proteins bind to the cytoplasmic domains of transmembranous glycoproteins. From the equimolar ratio of glycoproteins and nucleocapsid proteins in the mature virions it is suggested that each glycoprotein subunit interacts with a single capsid protein. As this association is repeated, the plasma membrane is wrapped around the nucleocapsid. From biochemical studies this process seems to be driven only by the sequential interactions of transmembranous glycoproteins E2 and E1 with capsid proteins.[16,19] Authors who favor this view of events assume that PE2 is cleaved to E2 prior to these events during transport in Golgi vesicles. Another school of thought holds that reorganization of at least PE2 in the cell membrane which at this point is not yet cleaved is a further step necessary for budding.[22] Based mainly on electron microscopic observations, it was suggested that after the first contact of a capsid protein with PE2, conformational changes occur on PE2 which allow cleavage to take place. In this case, energy necessary to drive the budding process to completion comes from both processes, binding and conformational changes. Evidence for conformational changes comes also from immunological studies using antibodies against viral envelope proteins.[48-50,54,55] The transitions in membrane morphology during budding suggest that processing of PE2 to E2 is a morphogenetic cleavage induced by binding of nucleocapsids to the membrane. Morphogenesis of flaviviruses was described by Westaway.[41] In contrast to alphaviruses, morphogenesis of flaviviruses is associated predominantly with membranes of the endoplasmatic reticulum and to a lesser extent with those of the Golgi complex. There are no convincing descriptions of preformed *Flavivirus* nucleocapsids. Morphologically mature virions are formed within the lumen of membranes of the Golgi complex.

XI. SIMILARITIES AND DIFFERENCES IN VERTEBRATE AND ARTHROPOD CELLS

The picture which emerges from comparison of growth characteristics of togaviruses in vertebrate and arthropod cells is that unlike vertebrate cells where *Togavirus* infections usually are short-term and cytocidal, infections of cultured mosquito cells by these viruses are long-term and inapparent. The virus yield in mosquito cell cultures may, under certain conditions, be the same as in vertebrate cells. Although a gross cytopathic effect is usually not apparent in mosquito cell cultures, some destructive processes may result from virus infection. In early experiments, virus titer obtained in mosquito cells was very low compared with that in vertebrate cells. The percentage of cells which were actually infected after incubation of the culture varied from 8 to 80%. In addition, virus yield in mosquito cells depended very much on the virus strain used for inoculation. These observations were extended by the finding that the yield of a definite virus strain varied considerably according to the cell clone used.[7] With carefully selected virus strains and cell clones, virtually all cells contained viral antigen during the initial (the acute) stage of infection.[56] Under those conditions, other characteristics of *Togavirus* replication in mosquito cells during this initial phase were also similar to that in vertebrate cells, with the exception of a slightly prolonged latent period in the poikilothermic cell culture if the optimum temperature for cell growth was used during viral growth. In this initial phase of infection, some togaviruses (mainly flaviviruses) exhibit sufficient cytopathic effect to form plaques.[57] Syncytium formation in *Aedes albopictus* cells induced by *Alphavirus* infection was reported by Späth and Koblet.[58] No qualitative differences were observed in the species of RNA and protein generated in mosquito cells as opposed to vertebrate cells. It is therefore improbable that the viral replication strategy differs in the two cell types. Following the initial phase, fundamental differences appear between the infection cycle in vertebrate and the infection of arthropod cells. At the time when peak titers of infectious virus are being released from infected mosquito cells, viral replication is reduced to a low level, but not completely eliminated. The resultant carrier cells can be subcultured for months or years. Often several types of virus particles are released from persistently infected mosquito cells with the following characteristics: small plaque variants, temperature-sensitive mutants, and defective interfering particles.[7,8,59,60] Characteristics of persistent infections of mosquito cells by togaviruses are described in the following article. Yet establishment of a persistent infection is not only possible in mosquito cells, it is also described in vertebrate cells, but only as a rare event and under specific conditions.[61,62]

Also in the initial phase virus replication in cultured mosquito cells and in vertebrate cells differs in a number of important respects. The bulk of mRNA in infected vertebrate cells is of viral origin. In mosquito cells there is no such substitution of cellular mRNA by viral mRNA. Therefore in contrast to vertebrate cells, it is difficult to characterize all viral specified proteins in infected mosquito cells because of the massive production of cellular proteins. (For comparison of viral effects on host macromolecular synthesis see the following section.)

A marked difference exists in the requirement of host functions for *Alphavirus* replication in vertebrate and mosquito cells. Virus maturation in mosquito cells is dependent on a constant production of some cellular functions which evidence was shown by actinomycin D sensitivity and host cell nuclear function of virus replication.[51,63]

It has been shown that the process of *Alphavirus* development in mosquito cells differs from that observed in cultured vertebrate cells. The budding of virions from the infected cell surface which is characteristic of infected vertebrate cells was rarely found in infected mosquito cell cultures. The bulk of virions produced in mosquito cells was detected in internally located vesicles of complex structure. Virions were released later by fusion of the vesicle with the cell surface. Viral nucleocapsids were found in the cytoplasmic vesicles in various stages of envelopment, suggesting that virus assembly could also take place in these internalized structures. In contrast to vertebrate cells,

Alphavirus development in mosquito cells occurs without massive accumulation of nucleocapsids and late in infection nucleocapsids are detected infrequently.[64-66] Gliedman et al.[64] suggested that the localization of some aspects of the virus reproductive process to restricted areas of the cell cytoplasm might protect the invertebrate cell from toxic effects that result from virus infection and cause the death of infected vertebrate cells. This idea would also be consistent with the observation of increased permeability of the plasma membrane in infected vertebrate cells[67] if it is concluded that alteration of permeability is caused by insertion of virus-specific proteins and their corresponding lipids into the plasma membranes. Lipids in the viral membrane are not a precise reflection of the cellular lipids. Viral envelope proteins play a role in selecting certain lipids for inclusion in the viral membrane. For example, membranes of alphaviruses grown in vertebrate cells had a higher microviscosity than that grown in mosquito cells.[68] The difference in microviscosity depends only on the lipids in the membrane of alphaviruses, since the viral proteins are similar whether the virus is grown in mosquito or in vertebrate cells.[69]

In the *Flavivirus* system, differences in morphogenesis between vertebrate and arthropod cells are not so dramatic. Mature virions are found in enlarged cisternae of the rough endoplasmatic reticulum membranes of mosquito cells.[70]

There are some minor differences in the composition of the virions replicated in the different cell systems. Since sialic acid is not found in cells of invertebrates, alphaviruses grown in mosquito cells were devoid of sialic acid.[7] A detailed study has been reported comparing the phospholipids of Semliki Forest virus grown in BHK cells with those in *Aedes albopictus* cells. Although the total amounts of phospholipids are similar in the viral membranes, notable quantitative differences are evident.[69] Antigenicity and infectivity of togaviruses grown in mosquito cells and vertebrate cells are similar.[7]

Thus, presence or absence of sialic acid as well as other host-dependent differences in viral envelope composition appear to be irrelevant with regard to antigenic specificity, biologic activity, and structural identity.

XII. EFFECTS OF VIRUS REPLICATION ON HOST CELL MACROMOLECULAR SYNTHESIS

A drastic inhibition of host-cell protein, RNA, and DNA synthesis takes place in vertebrate cells infected with alphaviruses (but only to a lesser extent with flaviviruses).[41] At the end of the exponential phase of virus growth, 70 to 90% of the newly synthesized RNA is of viral origin. It has been concluded, therefore, that virus replication leads to a significant reduction of host-cell RNA synthesis. There is only a limited number of experimental data on the host-cell DNA synthesis which suggests that DNA synthesis is also inhibited after infection with alphaviruses.[71] At the end of the exponential phase of virus growth the majority of the newly synthesized proteins are virus specific.[29,71] The mechanism of this inhibition is not yet clear. Several data suggest that the overall inhibition of protein synthesis is not reflected by a corresponding decrease in the number of ribosomes actively engaged in protein synthesis.[71] Competition between viral and host mRNA has been concluded to be the cause of host-cell inhibition. A viral mRNA (presumably 26S RNA) with a high affinity for initiation, but a slow rate of translation, could possibly explain both the mRNA substitution and reduction in the rate of overall protein synthesis.[71] Synthesis and accumulation of viral protein was stated to be the cause of host-cell inhibition.[22] Support for an indirect mechanism involving changes in the intracellular Na$^+$ and K$^+$ ion concentrations comes from observations that lowering or raising the Na$^+$ concentration in the culture fluid has little effect on the synthesis of *Alphavirus* proteins, but inhibit the synthesis of host-specified proteins in uninfected cells.[52,73,74] Garry et al.[75] demonstrated that alphaviruses inhibit translation of host-cell mRNA apparently by increasing the Na[+] concentration and lowering the K[+] concentration inside the cell. The viral messengers are translated efficiently under these altered conditions, whereas most host-cell mRNAs are not. The interference with translation is at the level of initiation. The viral structural proteins may be implicated in this inhibition[76] and it has been suggested that the altered ionic environment inside the cell results from interference with the Na[+]/K[+] pump.[77] A good deal of evidence has been accumulated demonstrating that animal cells lose permeability barriers upon infection with a variety of viruses.[67] On the basis of these data it has been suggested[78] that the specific inhibition of host-protein synthesis (shutoff phenomenon) induced by virus infection may be due to a general mechanism involving increased permeability at the plasma membranes with a consequent disturbance of the intracellular concentration of monovalent and divalent cations.[79-81] In this context it would be of interest to discuss the differences which exist in the replication of alphaviruses between arthropod and vertebrate cells. As it was suggested by Gliedman et al.,[64] virus assembly is sequestered in infected mosquito cells and possibly sequestering of the synthesis of virus-specific macromolecules occurs in *Alphavirus*-infected arthropod cells. By this localization of some aspects of the virus reproductive process to restricted areas within the cell cytoplasm, invertebrate cells might be protected from toxic agents that result from virus infection and cause the death of the infected vertebrate cell.

More comparative studies on the growth of togaviruses in arthropod and vertebrate cells will be necessary before it can be decided why mosquito cells seem to be relatively unimpaired by virus infection.

REFERENCES

1. **Mussgay, M.**, Growth cycle of arboviruses in vertebrate and arthropod cells, *Prog. Med. Virol.*, 6, 193, 1964.
2. **Singh, K. R. P.**, Cell cultures derived from larvae of *Aedes albopictus* (Skuse) and *Aedes aegypti* (L), *Curr. Sci.*, 36, 506, 1967.
3. **Vago, C.**, Invertebrate tissue culture, in *Methods in Virology*, Vol. 1, Maramorosch, K. and Koprowski, H., Eds., Academic Press, New York, 1967, 567.
4. **Vaughn, J. L.**, A review of the use of insect tissue culture for the study of insect-associated viruses, *Curr. Top. Microbiol. Immunol.*, 42, 102, 1968.
5. **Grace, T. D. C.**, Insect tissue culture and its use in virus research, *Adv. Virus Res.*, 14, 201, 1969.
6. **Singh, K. R. P.**, Growth of arboviruses in arthropod tissue culture, *Adv. Virus Res.*, 17, 187, 1972.
7. **Stollar, V.**, Togaviruses in cultured arthropod cells, in *The Togaviruses, Biology, Structure, Replication*, Schlesinger, R. W., Ed., Academic Press, New York, 1980, 584.
8. **Yunker, C. E.**, Arthropod tissue culture in the study of arboviruses and rickettsiae: a review, *Curr. Top. Microbiol. Immunol.*, 55, 113, 1971.
9. **Strauss, E. G.** and **Strauss, J. H.**, Replication strategies of the single stranded RNA viruses of eucaryotes, *Curr. Top. Microbiol. Immunol.*, 105, 1, 1983.
10. **Bachrach, H. L.**, Comparative strategies of animal virus replication, *Adv. Virus Res.*, 22, 163, 1974.
11. **Eaton, B. T.** and **Faulkner, P.**, Heterogeneity in the poly(A) content of the genome of Sindbis virus, *Virology*, 50, 865, 1972.
12. **Dobos, P.** and **Faulkner, P.**, Molecular weight of Sindbis virus ribonucleic acid as measured by polyacrylamide gel electrophoresis, *J. Virol.*, 6, 145, 1970.
13. **Simmons, D. T.** and **Strauss, J. H.**, Replication of Sindbis virus. I. Relative size and genetic content of 26S and 49S RNA, *J. Mol. Biol.*, 71, 599, 1972.
14. **Martin, B. A. B.** and **Burke, D. C.**, The replication of Semliki Forest virus, *J. Gen. Virol.*, 24, 45, 1974.
15. **Pfefferkorn, E. R.** and **Shapiro, D.**, Reproduction of togaviruses, in *Comprehensive Virology*, Vol. 2, Fraenkel-Conrat, H. and Wagner, R. R., Eds., Plenum Press, New York, 1974, 171.

16. Strauss, J. H. and Strauss, E. G., Togaviruses, in *The Molecular Biology of Animal Viruses*, Nayak, D. P., Ed., Marcel Dekker, New York, 1977, 111.

17. Wengler, G. and Wengler, G., Terminal sequences of the genome and replicative-form RNA of the flavivirus West Nile virus: absence of poly(A) and possible role in RNA replication, *Virology*, 113, 544, 1981.

18. Russel, P. K., Brandt, W. E., and Dalrymple, J. M., Chemical and antigenic structure of flaviviruses, in *The Togaviruses, Biology, Structure, Replication*, Schlesinger, R. W., Ed., Academic Press, New York, 1980, 501.

19. Garoff, H., Kondor-Koch, C., and Riedel, H., Structure and assembly of alphaviruses, *Curr. Top. Microbiol. Immunol.*, 99, 1, 1982.

20. Horzinek, M. C., The structure of togaviruses, *Prog. Med. Virol.*, 16, 109, 1973.

21. Murphy, F. A., Togavirus morphology and morphogenesis, in *The Togaviruses, Biology, Structure, Replication*, Schlesinger, R. W., Ed., Academic Press, New York, 1980, 241.

22. Brown, D. T., The assembly of alphaviruses, in *The Togaviruses, Biology, Structure, Replication*, Schlesinger, R. W., Ed., Academic Press, New York, 1980, 473.

23. Kääriäinen, L. and Söderlund, H., Structure and replication of alphaviruses, *Curr. Top. Microbiol. Immunol.*, 82, 15, 1978.

24. Enzmann, P.-J. and Weiland, F., Studies on the morphology of alphaviruses, *Virology*, 95, 501, 1979.

25. v. Bonsdorff, C.-H. and Harrison, S. C., Sindbis virus glycoproteins form a regular icosahedral surface lattice, *J. Virol.*, 16, 141, 1975.

26. Uterman, G. and Simons, K., Studies on the amphipatic nature of the membrane proteins in Semliki Forest virus, *J. Mol. Biol.*, 85, 569, 1974.

27. Garoff, H. and Simons, K., Location of the spike glycoproteins in the Semliki Forest virus membrane, *Proc. Natl. Acad. Sci. U.S.A.*, 71, 3988, 1974.

28. Söderlund, H., v. Bonsdorff, C.-H., and Ulmanen, I., Comparison of the structural properties of Sindbis and Semliki Forest virus nucleocapsids, *J. Gen. Virol.*, 45, 15, 1979.

29. Mussgay, M., Enzmann, P.-J., Horzinek, M., and Weiland, E., Growth cycle of arboviruses in vertebrate and arthropod cells, *Prog. Med. Virol.*, 19, 257, 1975.

30. Peleg, J. and Pecht, M., Adaptation of an *Aedes aegypti* mosquito cell line to growth at 15°C and its response to infection by Sindbis virus, *J. Gen. Virol.*, 38, 231, 1978.

31. Hahon, N. and Cooke, K. O., Primary virus-cell interactions in the immunofluorescence assay of VEE virus, *J. Virol.*, 1, 317, 1967.

32. Helenius, A., Morein, B., Fries, E., Simons, K., Robinson, P., Schirrmacher, V., Terhorst, C., and Strominger, J. L., Human (HLA-A and HLA-B) and murine (H-2K and H-2D) histocompatibility antigens are cell surface receptors for Semliki Forest virus, *Proc. Natl. Acad. Sci. U.S.A.*, 75, 3846, 1978.

33. Oldstone, M. B. A., Tishon, A., Dutko, F. J., Kennedy, S. I. T., Holland, J. J., and Lampert, P. W., Does the major histocompatibility complex serve as a specific receptor for Semliki Forest Virus?, *J. Virol.*, 34, 256, 1980.

34. Helenius, N., Kartenbeck, J., Simons, K., and Fries, E., On the entry of Semliki Forest virus into BHK-21 cells, *J. Cell Biol.*, 84, 404, 1980.

35. Fan, D. P. and Sefton, B. M., The entry into host cells of Sindbis virus, Vesicular Stomatitis virus and Sendai virus, *Cell*, 15, 985, 1978.

36. Morgan, C. and Howe, C., Structure and development of viruses as observed in the electron microscope. IX. Entry of parainfluenza (Sendai) virus, *J. Virol.*, 2, 1122, 1968.

37. Dales, S., Early events in cell-animal virus interactions, *Bacteriol. Rev.*, 37, 103, 1973.

38. Schlesinger, M. J. and Kääriäinen, L., Translation and processing of Alphavirus proteins, in *The Togaviruses, Biology, Structure, Replication*, Schlesinger, R. W., Ed., Academic Press, New York, 1980, 371.

39. Garoff, H. and Söderlund, H., The amphiphilic membrane glycoproteins of Semliki Forest virus are attached to the lipid bilayer by their COOH-terminal ends, *J. Mol. Biol.*, 124, 535, 1978.

40. Rice, C. M., Bell, J. R., Hunkapiller, M. W., Strauss, E. G., and Strauss, J. H., Isolation and characterization of the hydrophilic COOH-terminal domains of the Sindbis virion glycoproteins, *J. Mol. Biol.*, 154, 355, 1982.

41. Westaway, E. G., Replication of flaviviruses, in *The Togaviruses, Biology, Structure, Replication*, Schlesinger, R. W., Ed., Academic Press, New York, 1980, 531.

42. Strauss, E. G. and Strauss, J. H., Mutants of alphaviruses: genetics and physiology, in *The Togaviruses, Biology, Structure, Replication*, Schlesinger, R. W., Ed., Academic Press, New York, 1980, 393.

43. Frey, T. K., Gard, D. L., and Strauss, J. H., Biophysical studies on circle formation of Sindbis virus 49S RNA, *J. Mol. Biol.*, 132, 1, 1979.

44. **Kennedy, S. I. T.**, Synthesis of Alphavirus RNA, in *The Togaviruses, Biology, Structure, Replication*, Schlesinger, R. W., Ed., Academic Press, New York, 1980, 351.
45. **Garoff, H., Frischauf, A. M., Simons, K., Lehrbach, H., and Delius, H.**, Nucleotide sequence of cDNA coding for Semliki Forest virus membrane glycoproteins, *Nature*, 228, 235, 1980.
46. **Rice, C. M. and Strauss, J. H.**, Nucleotide sequence of the 26S RNA of Sindbis virus and deduced sequence of the encoded virus structural proteins, *Proc. Natl. Acad. Sci. U.S.A.*, 78, 2062, 1981.
47. **Bracha, M. and Schlesinger, M. J.**, Defects in RNA+ temperature-sensitive mutants of Sindbis virus and evidence for a complex of PE2-E1 viral glycoproteins, *Virology*, 74, 441, 1976.
48. **Smith, J. F. and Brown, T. T.**, Envelopment of Sindbis virus: synthesis and organization of proteins in cells infected with wild type and maturation defective mutants, *J. Virol.*, 22, 662, 1977.
49. **Kaluza, G., Rott, R., and Schwarz, R.**, Carbohydrate-induced conformational changes of Semliki Forest virus glycoproteins determine antigenicity, *Virology*, 102, 286, 1980.
50. **Kaluza, G. and Pauli, G.**, The influence of intramolecular disulfide bonds on the structure and function of Semliki Forest virus membrane glycoproteins, *Virology*, 102, 300, 1980.
51. **Scheefers-Borchel, U., Scheefers, H., Edwards, J., and Brown, D. T.**, Sindbis virus maturation in cultured mosquito cells is sensitive to actinomycin D, *Virology*, 110, 292, 1981.
52. **Bell, J. W., Garry, R. F., and Waite, M. R. F.**, Effect of low NaCl medium on the envelope glycoproteins of Sindbis virus, *J. Virol.*, 25, 764, 1978.
53. **Compans, R. W. and Klenk, H.-D.**, Viral membranes, in *Comprehensive Virology*, Vol. 13, Fraenkel-Conrat, H. and Wagner, R., Eds., Plenum Press, New York, 1979, 293.
54. **Bell, J. W. and Waite, M. R. F.**, Envelope antigens of Sindbis virus in cells infected with temperature-sensitive mutants, *J. Virol.*, 21, 788, 1977.
55. **Rice, C. M. and Strauss, J. H.**, Association of Sindbis virus glycoproteins and their precursors, *J. Mol. Biol.*, 154, 325, 1982.
56. **Igarashi, A., Koo, R., and Stollar, V.**, Evolution and properties of *Aedes albopictus* cell cultures persistently infected with Sindbis virus, *Virology*, 82, 69, 1977.
57. **Yunker, C. E. and Cory, J.**, Plaque production by arboviruses in Singh's *Aedes albopictus* cells, *Appl. Microbiol.*, 29, 81, 1975.
58. **Späth, P. J. and Koblet, H.**, Alphavirus induced syncytium formation in *Aedes albopictus* cell cultures, in *Invertebrate Systems In Vitro*, Kurstak, E., Maramorosch, K., and Dübendorfer, F., Eds., Elsevier/North-Holland, Amsterdam, 1980, 375.
59. **Schlesinger, R. W., Stollar, V., Igarashi, A., Guild, G. M., and Cleaves, G.**, Natural life cycle of arthropod-borne Togaviruses: inference from cell culture models, in *Viruses and Environment*, Kurstak, E. and Maramorosch, K., Eds., Academic Press, New York, 1978, 281.
60. **Eaton, B. T.**, Persistent Togavirus infection of *Aedes albopictus* cells, in *Viruses and Environment*, Kurstak, E. and Maramorosch, K., Eds., Academic Press, New York, 1978, 181.
61. **Schwöbel, W. and Ahl, R.**, Persistence of Sindbis virus in BHK-21 cell cultures, *Arch. Gesamte Virusforsch.*, 38, 1, 1972.
62. **Inglot, A. D., Albin, M., and Chudzio, T.**, Persistent infections of mouse cells with Sindbis virus: role of virulence of strains, autointerfering particles and interferon, *J. Gen. Virol.*, 20, 105, 1973.
63. **Erwin, C. and Brown, D. T.**, Requirement of cell nucleus for Sindbis virus replication in cultured *Aedes albopictus* cells, *J. Virol.*, 45, 792, 1983.
64. **Gliedman, J. B., Smith, J. F., and Brown, D. T.**, Morphogenesis of Sindbis virus in cultured *Aedes albopictus* cells, *J. Virol.*, 16, 913, 1975.
65. **Raghow, R. S., Davey, M. W., and Dalgarno, L.**, The growth of Semliki Forest virus in cultured mosquito cells: ultrastructural observations, *Arch. Gesamte Virusforsch.*, 43, 165, 1973a.
66. **Raghow, R. S., Grace, T. D. C., Filshie, B. K., Bartley, W., and Dalgarno, L.**, Ross River virus multiplication in cultured mosquito and mammalian cells: virus growth and correlated ultrastructural changes, *J. Gen. Virol.*, 21, 109, 1973b.
67. **Kohn, A.**, Early interactions of viruses with cellular membranes, *Adv. Virus Res.*, 24, 223, 1979.
68. **Moore, N. F., Barenholz, Y., and Wagner, R. R.**, Microviscosity of Togavirus membranes studied by fluorescent depolarization: influence of envelope proteins and the host cell, *J. Virol.*, 19, 126, 1976.
69. **Luukkonen, A., v. Bonsdorff, C.-H., and Renkonen, O.**, Characterization of Semliki Forest virus grown in mosquito cells. Comparison with the virus from hamster cells, *Virology*, 78, 331, 1977.
70. **Ko, K. K., Igarashi, A., and Fukai, K.**, Electron microscopic observations on *Aedes albopictus* cells infected with Dengue viruses, *Arch. Virol.*, 62, 41, 1979.
71. **Wengler, G.**, Effects of alphaviruses on host cell macromolecular synthesis, in *The Togaviruses, Biology, Structure, Replication*, Schlesinger, R. W., Ed., Academic Press, New York, 1980, 459.
72. **Mussgay, M., Enzmann, P.-J., and Horst, J.**, Influence of an arbovirus infection (Sindbis virus) on the protein and ribonucleic acid synthesis of cultivated chick embryo cells, *Arch. Gesamte Virusforsch.*, 31, 81, 1970.

73. Waite, M. R. F. and Pfefferkorn, E. R., The effect of altered osmotic pressure on the growth of Sindbis virus, *J. Virol.*, 2, 759, 1968.
74. Clegg, J. C. S. and Kennedy, S. I. T., Initiation of synthesis of the structural proteins of Semliki Forest virus, *J. Mol. Biol.*, 97, 401, 1975.
75. Garry, R. F., Bishop, J. M., Parker, S., Westbrook, K., Lewis, G., and Waite, M. R. F., Na⁺ and K⁺ concentrations and the regulation of protein synthesis in Sindbis virus infected chick cells, *Virology*, 96, 108, 1979a.
76. Atkins, G. J., The effect of infection with Sindbis virus and its temperature-sensitive mutants on cellular protein and DNA synthesis, *Virology*, 71, 593, 1976.
77. Garry, R. F., Westbrook, K., and Waite, M. R. F., Differential effects of ouabain on host- and Sindbis virus-specific protein synthesis, *Virology*, 99, 179, 1979b.
78. Carrasco, L., The development of new agents based on virus-mediated cell modification, in *Antiviral Mechanisms for the Control of Neoplasia*, Chanara, P., Ed., Plenum Press, New York, 1979, 623.
79. Carrasco, L., Membrane leakiness after viral infection and a new approach to the development of antiviral agents, *Nature (London)*, 272, 694, 1978.
80. Durham, A. C. H., Do viruses use calcium ions to shut-off host cell functions?, *Nature (London)*, 267, 375, 1977.

Chapter 16

PERSISTENT INFECTIONS OF MOSQUITO CELLS BY TOGAVIRUSES

P.-J. Enzmann

TABLE OF CONTENTS

I. Introduction...68

II. Mechanisms of Virus Persistence 68
    A. Transformation of Cells by Viral Nucleic Acid............................68
    B. Defective Interfering Virus Particles 68
    C. Temperature-Sensitive Mutants 69
    D. Antiviral Substances 69

III. Cytolytic Infections of Mosquito Cells 69

IV. Kinetics of Persistence 70

V. Viral Functions in Persistently Infected Cells 71
    A. Protein Synthesis in Infected Cells...71
    B. Viral RNA Synthesis 72
    C. Temperature Sensitivity 72

VI. Characteristics of Persistently Infected Cells...................................73

VII. Interpretation of Persistence 73

References 75

I. INTRODUCTION

In a persistent infection the virus either continues to multiply and remains demonstrable or it may survive in some noninfectious form with the potential to be reactivated later. Persistent viral infections can be characterized by low or high levels of virus production or an absence of complete virus maturation, but with maintenance of either viral genomes within infected cells or limited expression of viral antigens.

For convenience, persistent infections can be divided into three categories as proposed by Fenner,[1] recognizing that there is some overlap:

Latent infections — Persistent infections with intermittent acute episodes of disease between which virus is usually not demonstrable
Chronic infections — Persistent infections in which the virus is always demonstrable and often shed, but disease is either absent or is associated with immunopathological disturbances
Slow infections — Persistent infections with a long incubation period followed by a slowly progressive disease that is usually lethal

These definitions were made regarding infected organisms. For persistently infected cells in culture these terms may be adapted. Persistent infections of cultured cells with viruses can therefore be divided into three types: transformation of cells by viruses, carrier cultures, and steady state infections. As mentioned above, this division is not rigid because there is some overlap. Viral persistence in cultured cells, especially in cultured mosquito cells, is the topic of this chapter. References are selective, rather than comprehensive, review articles are cited whenever possible and primary articles are included when necessary.

II. MECHANISMS OF VIRUS PERSISTENCE

A. Transformation of Cells by Viral Nucleic Acid

Retroviridae have been unequivocally associated with neoplastic disease, but their importance is heightened by the fact that they possess an enzyme (reverse transcriptase) which catalyzes the synthesis of a DNA complementary to the viral RNA genome. The virus-specific DNA is then integrated within the host nucleic acid. Tumor viruses establish this type of persistent infection.[2] Togaviruses do not possess a reverse transcriptase, but it was reported by Zhdanov[3] that the Sindbis virus and tick-borne encephalitis (TBE) virus can use the enzyme of a naturally occurring *Retrovirus* or cellular RNA-dependent DNA polymerase to integrate proviral DNA into the cellular genome. One can speculate that certain ecological processes may be related to viral genome integration. However, these results could not be confirmed by others in vertebrate as well as in *Aedes albopictus* cells.[4,5] The phenomenon of persistence of togaviruses in mosquito cells cannot be, as a rule, attributed to the integration of proviral DNA into host DNA. Possibly, but only in special cases, this process may occur in *Togavirus* persistence. The main attention in this respect is called to the mechanisms which are described as follows.

B. Defective Interfering Virus Particles

One main mechanism which is involved in the establishment and maintenance of persistence is the generation of defective interfering (DI) particles. A DI particle is defined as a virus which is defective in that it cannot autonomously replicate unless the cell is coinfected with a standard virus. In such a mixed infection with both normal infectious virions and DI particles of the same virus strain, DI particles multiply and interfere with the replication of the complete virus.[6]

C. Temperature-Sensitive Mutants

Evidence for the involvement of temperature-sensitive (ts) mutants in persistent infection is derived from two types of experiments. In a number of studies it has been found that the virus released from persistently infected cells into the medium contains ts mutants. In other studies, growth of persistently infected cells at lower incubation temperature led to an increase in the amount of extracellular virus which in some cases appeared to be temperature-sensitive.[7] Schwöbel and Ahl[8] isolated a small plaque variant from BHK-21 cells persistently infected with the Sindbis virus, but they did not report on the temperature-sensitivity of the strain. Stollar et al.[9] and Shenk et al.[10] reported the isolation and purification of small plaque mutants of persistent infections of the Sindbis virus in *Aedes aegypti* and *A. albopictus* cells, respectively. The small plaque mutant isolated from *A. aegypti* cells was shown to be an RNA$^+$ ts mutant and interfered with the replication of the wild type virus at the permissive and nonpermissive temperature. Plaque morphology and temperature-sensitivity appeared to be independent characters since temperature-sensitive large-plaque variants from original small-plaque WEE virus could be isolated from where small-plaque revertants emerged.[11]

D. Antiviral Substances

Establishment of antiviral activity occurs after several hours of cell exposure to interferon. This type of antiviral factor is induced in cells infected with both infective and inactivated viruses. Interferon production usually continues for several hours. It then stops and can be reinitiated only after a refractory period. Endogenous production of antiviral activity may play a role in maintaining persistent infections.[12] There are several reports on the appearance of antiviral substances in mosquito cells infected with togaviruses, yet these reports are at least partially contradictory (see Section VI).

The generation of DI particles and ts mutants and their involvement in persistent infections, as well as the action of the antiviral substances in this respect will be discussed in detail later.

III. CYTOLYTIC INFECTIONS OF MOSQUITO CELLS

Arthropod vectors as a rule show no pathological effects upon infection by togaviruses, while naturally and experimentally infected vertebrates often do. These facts seemed to be confirmed in cell culture systems derived from arthropods and vertebrates, especially for alphaviruses. Infection is highly cytolytic for permissive vertebrate-cell cultures, while permissive arthropod cell cultures display no detectable cytopathic effects (CPE); normally they become persistently infected. This contrast between cytolytic infection of vertebrate cells and noncytolytic infection of arthropod cells is not absolute, since it is possible to establish chronically infected vertebrate cell cultures[8,13] and, conversely, some destructive processes may result from a virus infection in arthropod cells depending on several conditions for infection. In early experiments, virus titer obtained in mosquito cells was very low compared with that in vertebrate cells. The percentage of cells which were actually infected after incubation of the culture varied from 8 to 80%. In addition, the virus yield in mosquito cells depended very much on the virus strain used for inoculation. These observations were extended by the finding that the yield of a definite virus strain varied considerably according to the cell clone used.[14] A similar observation could be made regarding the appearance of a CPE in mosquito cells after infection with togaviruses. Cloned variants of *Aedes albopictus* cells differ markedly in their response to infection with the Sindbis virus.[14] Several clones showed no CPE and others manifested a severe CPE in less than 20 hr after infection. The appearance of CPE seemed to also be dependent on the cell-culture medium. An additional effect is given by dependence of CPE on temperature. At 28°C, defined cell-clones showed no signs of CPE, whereas at 34 to 37°C there was clear evidence of cell destruction. There was no convincing correlation between the amounts of virus produced and the occurrence of CPE. In these clones, viral not cellular, RNA synthesis is increased by raising the incubation temperature from 28 to 34 to 37°C. Thus, both viral RNA synthesis and CPE showed an important dependence on temperature, but the association between increased viral RNA synthesis and the occurrence of CPE does not necessarily indicate a cause-effect relationship. In the first reports of arboviruses producing CPE in arthropod cell cultures, all the viruses which act in such a way belonged to the *Flavivirus* group. CPE was not observed after infection with alphaviruses. The main features of CPE with those viruses are cytolysis of individual cells, development of large syncytial masses, gradual increase in the number of multinucleated giant cells, phagocytosis of dead cells, and ultimate recovery of the infected cultures. In the meantime, CPE was also reported in cultures infected with alphaviruses under certain circumstances as described above. In addition, syncytium formation of mosquito cells induced by alphaviruses was described recently which reflects many details of the normal reaction of mosquito cells to *Flavivirus* replication. As a rule, signs of CPE are visible only in the initial, i.e., acute phase of infection. Even in those instances when CPE can be demonstrated in *Togavirus*-infected mosquito cells, the cells recover to yield a persistently infected culture.[14,16]

IV. KINETICS OF PERSISTENCE

In the initial phase of infection, mosquito cells replicate togaviruses in a similar way as do vertebrate cells. In mosquito cells, there is a latent period of 5 to 6 hr for alphaviruses (this is somewhat longer for flaviviruses) after which virus production increases rapidly and virus yields plateau after 14 to 24 hrs (see Chapter 15, this volume). Little or no CPE was observed in early experiments in this phase as described above. After the initial burst of virus production, the titer declines rapidly at the first week and then more slowly over several weeks and then it finally plateaus. The resultant persistently infected culture is stable and can be maintained over long periods. Usually it was thought that the persistent infection is established in the infected mosquito cell culture without loss of cells from the culture or any obvious alteration in any of the growth characteristics of the infected culture.

The early experiments were performed with uncloned cultures of mosquito cells. As described above, cloned variants differ in their response to infection with alphaviruses.[14] The percentage of cells which were actually infected in a persistently infected culture varied from less than 1%[17,18] to 40 to 80%.[4] During the first week after infection of *Aedes albopictus* cells with Sindbis virus viral titers ranged between 10^8 and 10^9 PFU. Virus titers then dropped and within a few weeks stabilized at about 10^5 to 10^6 PFU. After subculturing of persistently infected cultures, viral yields were maximal during the first 3 to 4 days. Thereafter, although the cell number continued to increase for several days, viral yields decreased sharply to the level before subculturing.[4] In another study a cloned line of *A. albopictus* cells infected with CHIK virus gave yields around 10^8 PFU about 24 hr after infection.[19] After infection of cloned *A. albopictus* cells with Ross River virus and Semliki Forest virus, maximum virus titers were reached at 34 to 48 hr. At 12 hr post-infection (p.i.), 85% of cells assayed as infected by infective center assay; by 48 hr p.i. persistence was established and at this time virus production was reduced and only 5% of cells assayed as infected.[20] Tooker and Kennedy[21] reported also on the infection of cloned *A. albopictus* cells with Semliki Forest virus. Virus yields, cytopathology, and kinetics of viral growth, as well as kinetics and mechanisms of development of persistence, varied in different clones. In 115 clonally purified *A. albopictus* cells virus yields fell into two nonoverlapping groups. The first group comprising 78 of the cell lines gave yields of between 8×10^6 and 2×10^8 PFU. The second group comprising 34 of the cell lines gave yields of between 1×10^9 and 8×10^9 PFU. Maximum virus titers were reached in the first group, the low-virus-producing lines, at about 20 hr after infection, and in the second group, the high-virus-producing lines, at about 15 hr p.i. All but three of the lines belonging to the first group exhibited no discernible CPE. All of the high-virus-producing lines showed moderate to severe CPE within 30 hr p.i.

An example for the development of persistence of a *Flavivirus* (Dengue 2) in *A. albopictus* cells was given by Sinarachatanant and Olson.[22] Maximal virus titers were reached at day 4 p.i. and declined gradually from approximately 10^6 to 10^7 PFU to about 10^4 PFU after 6 months of culture. Kuno[23] reported on the fluctuation of extracellular virus titer (Dengue 1, 2, 3) in persistently infected cultures of a nonvector mosquito cell line between 10^2 and 10^7 PFU. Ng and Westaway[24] reported on maximal Kunjin virus titers on day 2 after infection of *A. albopictus* cells.

Persistent infection of mosquito cells can be maintained over a long period, but mosquito cells also have the potency of "self-curing", a process which leads to virus-free cells. In the culture, virus-free cells may be reinfected by the virus in the medium in order to perpetuate persistently infected cultures. The process of self-curing may also depend on the virus strain since it seems to be easier to get virus-free cultures after infection with CHIK virus than with Sindbis virus.[4,19] Therefore the mechanism of *Togavirus* persistence in mosquito cells must be regarded not only on the level of the persistently infected culture, but also on the level of the single cell. The persistent phase of infection was found to be dependent upon the presence of extracellular virus for its maintenance. Prevention of reinfection of the cells by immune serum resulted in a rapid curing of the culture.[25]

V. VIRAL FUNCTIONS IN PERSISTENTLY INFECTED CELLS

A. Protein Synthesis in Infected Cells

Characteristics of *Togavirus* replication in mosquito cells in the acute (that is the initial) phase of infection are similar to that in vertebrate cells (see Chapter 15 in this volume). The same spectrum of structural and nonstructural viral proteins is found in infected mosquito cells as in vertebrate cells, but there is a marked difference in the quantity of viral proteins synthesized in mosquito cells as compared with vertebrate cells. Only the C-protein (nucleocapsid) of alphaviruses could readily be seen in infected mosquito cells (and also the E1 protein was seen by other authors). The other structural and nonstructural proteins could be identified after subtraction of a host-specific background after electrophoresis in polyacrylamide gels.[20,21,24] There is no accumulation of viral proteins in the cytoplasm of infected mosquito cells. The small amounts of virus-specified proteins contrast with the high yields of virus during the acute phase of infection. From these results an efficient mechanism for utilization of viral proteins in mosquito cells can be suggested. During the establishment of persistence there is a reduction in the synthesis of viral polypeptides which parallels the decrease in virus yield. After subculturing persistently infected cells there is a short increase in viral protein synthesis as well as in virus yield. Identification of viral proteins in infected mosquito cells is complicated by the fact that host protein synthesis is not inhibited, as it is a normal event in vertebrate cells.[20,24]

In a detailed study, proteins specified by Semliki Forest virus in infected *Aedes albopictus* cell clones were analyzed.[21] Of the six clones used, four gave a high virus yield (group 1) and showed CPE after infection. The remaining two clones gave low virus yield (group 2) without any signs of CPE. In contrast to the studies described above, all virus-specified proteins were detectable in group 1 lines, corresponding to the high virus yield in it; in addition, cellular protein synthesis was markedly reduced. A low level of protein synthesis corresponding to a low yield of virus and no shutoff of cellular protein synthesis was found in infected group 2 lines. There were no data on the percentage of cells which were infectable or which were actually infected in both groups of cell lines. Reigel and Koblet[26] reported on two proteins normally secreted into the medium by uncloned *A. albopictus* cells and by an uninfected clone of these cells which gave a high virus yield after infection with togaviruses. Both proteins disappeared upon infection of the high virus-producing clone during the time of virus burst, and reappeared when the cells changed to the permanently infected state. These secreted proteins were not affected in the uncloned cell line. The biological significance of this effect is not yet clear.

B. Viral RNA Synthesis

The replication of viral RNA in mosquito cells was reviewed by Eaton.[27] In these cells, the replication strategy is essentially the same as in vertebrate cells (see Chapter 15 in this volume). There is some evidence that the molar ratio of 42S RNA to 26S RNA in the initial phase of infection is not the same in mosquito cells as compared to vertebrate cells. The amount of 26S RNA in mosquito cells is much smaller than that of 42S RNA.[27] By using cloned variants of *A. albopictus* cells, Tooker and Kennedy[21] found no difference in the molar ratio of 42S RNA and 26S RNA.

Clones yielding low virus synthesized only small amounts of single-stranded viral RNA compared with clones that yielded a lot of virus, but as stated above, there were no data on the percentage of cells which were actually infected. Soon after the initial virus burst, viral RNA synthesis is severely restricted. This inhibition of viral RNA replication represents the first demonstrable step in the development of the persistent infection.[14,27,28] Although *Alphavirus* DI-particles are readily produced in vertebrate cells, serially undiluted passage of the Sindbis virus every 16 hr to *A. albopictus* cells does not generate DI-particles.[6,27,29] However, if mosquito cells are infected with virus preparations containing large amounts of DI-particles, synthesis of DI-RNA — as it occurs in vertebrate cells — is restricted in mosquito cells.[30] In spite of the failure to generate DI-particles by serial undiluted passage in the initial phase of virus infection, there is evidence that a few DI-particles are produced by *A. albopictus* cells when they are persistently infected.[27] However, plus-stranded RNA smaller than 26S RNA was found in mosquito cells between 7 and 13 hr after infection with alphaviruses.[21,27] This means that DI-RNA is produced when infected cells convert from the acute phase of infection to the persistent state of infection. Although defective RNAs represent the major part of viral RNA in persistently infected mosquito cells, they are very rarely assembled into virions.[29]

C. Temperature Sensitivity

Evolution and properties of ts viruses and small-plaque mutants after infection of mosquito cells with togaviruses have been reviewed by Stollar[14] and Eaton.[27] Such virus produces plaques at 28 or 34°C but not at higher temperatures (39°C). The virus population of persistently infected cultures shifted to the ts phenotype over a period of several weeks. The time required for small-plaque or ts viruses to replace standard or wild type viruses varies according to the virus or virus strain.[19] Most of the virus clones derived from Sindbis-virus infected *A. albopictus* cells are RNA⁺ and fail to complement each other.[10] The fact that the ts clones give rise to ts⁺ revertants that retain their small-plaque morphology indicates that the small-plaque ts variants contain at least two genetic lesions. Functional defects of the *Alphavirus* ts mutants are reviewed by Strauss and Strauss.[31] As demonstrated in the study of Igarashi et al.,[4] the virus population of persistently infected cultures shifted to the ts phenotype over a period of several weeks. The ts mutants readily complement in vertebrate cells, but fail to do so in mosquito cells.[32] This effect may be explained by the fact that virus synthesis in mosquito cells is sequestered (see Chapter 15 in this volume).

VI. CHARACTERISTICS OF PERSISTENTLY INFECTED CELLS

As already noted, togaviruses replicate in mosquito cells at about the same rate and to comparable titers as in vertebrate cells. Mosquito cells generally show no CPE with the exception of the system using cloned variants of the *A. albopictus* cell line. Even in those instances when CPE can be demonstrated in *Togavirus*-infected mosquito cells, the cells recover. After the initial burst of virus replication, mosquito cells enter into a state of virus persistence with low virus production, and usually small-plaque variants and temperature-sensitive mutants of the original virus are produced. However, for reasons not yet understood fully, mosquito cells only rarely generate DI-particles, as compared with vertebrate cells. Only DI-RNA is produced upon infection with a normal virus preparation containing no DI-particles. Davey and Dalgarno[28] found a limited but marked stimulation of viral replication upon subculturing persistently infected cells. There seems to be a mechanism working in mosquito cells for regulation of the synthesis of viral macromolecules. Curing of individual cells and also of cell cultures may result from this mechanism. Persistently infected cells show a high degree of resistance upon infection with normal, wild type virus.

For example, mosquito cells persistently infected with Sindbis virus are resistant to superinfection with the homologous virus, but can be infected by the EEE virus.[17] Both DI-RNA and ts mutants may be responsible for this homologous resistance. Eaton[27] suggested that the ability of persistently infected cells to resist superinfection with the homologous virus is a manifestation of the control exerted in these cells over the replication of the persisting virus. Besides homologous interference, a heterologous interference also exists in infected mosquito cells. Infected cultures were resistant to superinfection with a heterologous virus for approximately 8 days. After that time, infected cultures supported the replication of the heterologous virus to the same extent as did persistently infected cultures established months previously.[33] Persistent infection is initiated with the beginning of virus inhibition after the acute phase of infection.

Several factors are deemed responsible for this step. Enzmann[34] has reported data suggesting that interferon-like substances may be the reason for the inhibition of virus replication. However, other investigators have failed to find evidence for such products in infected mosquito cells.[25,28,36,37] Riedel and Brown[38] reported on a novel antiviral activity in the medium of persistently infected mosquito cells which differs from interferon in that it is virus-specific as well as cell-specific. The antiviral activity appeared in the medium at 48 to 72 hr after infection with the Sindbis virus. It remains unclear whether this substance is of viral or cellular origin. Evidence for secretion of an antiviral factor from mosquito cells infected with Banzi virus (genus *Flavivirus*) which is virus specific was presented by Lee and Schloemer.[39] This factor seems to be of viral origin since antiviral activity is blocked by treatment with anti-Banzi viral serum. Possibly this factor is related to a structural protein. Reduced dengue virus titers in cells pretreated with medium from persistently infected mosquito cells was reported by Igarashi.[19] In addition, infection of mosquito cells with one type of a ts dengue virus rendered the cells resistant to all types of standard dengue viruses.

VII. INTERPRETATION OF PERSISTENCE

Several mechanisms which are not necessarily exclusive have been proposed to explain the development of persistent infections in cultured mosquito cells. It is far from clear as to whether these mechanisms operate alone or together. One mechanism involves the development of defective viral RNA which interferes with replication of the parent viral genome by competition for a limited viral protein such as the replicase and transcriptase. Another mechanism focuses on the selection of ts and small-plaque viral mutants during the establishment of persistent infections of cultured cells which mutants suppress replication of the wild-type virus. Establishment of antiviral activity mediated by a protein is a third mechanism which is possibly responsible for limited viral macromolecular synthesis.

For interpretation of persistence it is necessary to recognize two aspects, persistent infection in an individual cell and persistent infection of a cell culture. Studies with mosquito cells cloned before, as well as after establishment of virus persistence, have shed some light onto the dynamics of persistent infections.[4,21,29] The mechanism of self-curing seems to work efficiently in *Togavirus*-infected mosquito cells. An individual cell previously infected but after self-curing free of virus may be reinfected and replicate virus in a manner comparable to cells without a history of previous infection. From this point of view the mechanism of self-curing seems to be regulated by cellular factors rather than by viral functions. Individual cells may differ in their ability to replicate virus and to eliminate the virus after a period of time, but this seems to be more of a quantitative problem than a qualitative one. Therefore from such cells cultures may arise which give high virus yield while others give low virus yield. Both types of cultures possess the ability to establish persistent infections. It is unclear, however, whether both types of cells possess the same ability to eliminate virus, for maintenance of a persistent infection in a cell culture reinfection of those cells that are virus-free is necessary.[25] Therefore the process for elimination of the virus from a cell culture is very prolonged compared with the process in individual cells. During this time ts-mutants arise which help to maintain the persistent state. The phenomena of homologous (and also in several cases heterologous) interference remain to be explained. From the course of time of appearance, the interference phenomenon seems to be linked with the restriction of virus replication which leads to self-curing. It is likely, therefore, that there exists a mechanism in mosquito cells for regulating the synthesis of viral macromolecules and ultimately of infectious virus progeny. The ability of an infected individual cell to resist superinfection may be caused by three mechanisms which need not necessarily be connected. As a first observation, DI-RNA appears a short time after infection of a cell. Antiviral factors and decreased RNA-synthesis are described in a later phase when infection of the culture enters the persistent state; neither phenomenon is attributable to an individual cell, but only to the culture since antiviral factors may act after secretion into cell culture medium and RNA-synthesis cannot be measured on the level of a single cell.

Antiviral agents and decreased RNA-synthesis may be involved in the establishment of persistence at a stage somewhat later than DI-RNA, but no conclusive evidence has been presented in this regard. The generation of DI-RNA before the establishment of a persistent infection and the continuous presence of this type of RNA in persistently infected cells supports the idea that interference by defective viral RNA with standard RNA may be involved in establishment and partially in the maintenance of a persistent infection of togaviruses in mosquito cells. For the establishment of persistence, the mechanisms of self-curing and interference can be regarded as related and may be caused by DI-RNA. For the maintenance of persistence, both mechanisms must be regarded separately since self-curing is directed against the virus presence. When a persistent infection is established in mosquito cells it seems that the nonpermissiveness of those cultures for superinfection with wild-type virus is best explained by assuming that the virus from cultures themselves (possibly ts-mutant) is the interfering agent.

For interpretation of virus persistence in mosquito cells it is worthwhile to look separately in individual cells and in the cell culture as a whole. It is further necessary to distinguish between establishment and maintenance of a persistent infection.

It is plausible that rather than a single mechanism being responsible, there are several factors which act simultaneously and successively to bring the system of persistent infection into balance and to hold it in balance since disturbing factors lead to self-curing.

Finally, it should be mentioned that persistent infections of mosquito cells with togaviruses seem to be an efficient, cell-dependent mechanism that allows these cells to escape the damaging effect of virus replication as it is seen in vertebrate cells. Since there are only minor differences in the replication strategy of the viral genome in arthropod and vertebrate cells (see Chapter 15 in this volume) the efficient establishment of a persistent infection in mosquito cells as compared to the very rare but demonstrable event in vertebrate cells, seems to be more of a quantitative problem than a qualitative one.

REFERENCES

1. Fenner, F. and White, D. O., Persistent infections, in *Medical Virology,* Fenner, F. and White, D. O., Eds., Academic Press, New York, 1976, 140.
2. Weinberg, R. A. and Steffen, D. L., Regulation of expression of the integrated Retrovirus genome, *J. Gen. Virol.*, 54, 1, 1981.
3. Zhdanov, V. M., Integration of genomes of infectious RNA viruses, *Intervirology*, 6, 129, 1976.
4. Igarashi, A., Koo, R., and Stollar, V., Evolution of properties of *Aedes albopictus* cell cultures persistently infected with Sindbis virus, *Virology*, 82, 69, 1977.
5. Streissle, G., Persistent viral infections as models for research in virus chemotherapy, *Adv. Virus Res.*, 26, 37, 1981.
6. Stollar, V., Defective interfering alphaviruses, in *The Togaviruses, Biology, Structure, Replication*, Schlesinger, R. W., Ed., Academic Press, New York, 1980, 427.
7. Rima, B. K. and Martin, S. J., Persistent infections of tissue culture cells by RNA viruses, *Med. Microbiol. Immunol.*, 162, 89, 1976.
8. Schwöbel, W. and Ahl, R., Persistence of Sindbis virus in BHK-21 cell cultures, *Arch. Gesamte Virusforsch.*, 38, 1, 1972.
9. Stollar, V., Peleg, J., and Shenk, T. E., Temperature sensitivity of Sindbis virus mutant isolated from persistently infected *Aedes aegypti* cell cultures, *Intervirology*, 2, 337, 1974.
10. Shenk, T. E., Koshelnyk, K. A., and Stollar, V., Temperature-sensitive virus from *Aedes albopictus* cells chronically infected with Sindbis virus, *J. Virol.*, 13, 439, 1974.
11. Simizu, B. and Takayama, N., Relationship between neurovirulence and temperature-sensitivity of an attenuated Western Equine Encephalitis virus, *Arch. Gesamte Virusforsch.*, 34, 242, 1971.
12. Friedman, R. M. and Ramseur, J. M., Mechanism of persistent infections by cytopathic viruses in tissue culture, *Arch. Virol.*, 60, 83, 1979.
13. Inglot, A. D., Albin, M., and Chudzio, T., Persistent infections of mouse cells with Sindbis virus: role of virulence of strains, autointerfering particles and interferon, *J. Gen. Virol.*, 20, 105, 1973.
14. Stollar, V., Togaviruses in cultured arthropod cells, in *The Togaviruses, Biology, Structure, Replication*, Schlesinger, R. W., Ed., Academic Press, New York, 1980, 584.
15. Singh, K. R. P., Growth of arboviruses in arthropod tissue culture, *Adv. Virus Res.*, 17, 187, 1972.
16. Späth, P. J. and Koblet, H., Alphavirus induced syncytium formation in *Aedes albopictus* cell cultures, in *Invertebrate Systems In Vitro*, Kurstak, E., Maramorosch, K., and Dübendorfer, A., Eds., Elsevier/North-Holland, Amsterdam, 1980, 375.
17. Peleg, J., Studies on the behavior of arboviruses in an *Aedes aegypti* mosquito cell line (Peleg), *Arch. Gesamte Virusforsch.*, 37, 54, 1972.
18. Davey, M. W., Dennet, D. P., and Dalgarno, L., The growth of two togaviruses in cultured mosquito and vertebrate cells, *J. Gen. Virol.*, 20, 225, 1973.
19. Igarashi, A., Production of temperature-sensitive and pathogenic virus from *Aedes albopictus* cells (Singh) presistently infected with Chikungunya virus, *Arch. Virol.*, 62, 303, 1979.

20. Richardson, M. A., Boulton, R. W., Raghow, R. S., and Dalgarno, L., Polypeptide synthesis in Alphavirus-infected *Aedes albopictus* cells during establishment of persistent infection, *Arch. Virol.*, 63, 263, 1980.
21. Tooker, P. and Kennedy, S. I. T., Semliki Forest virus multiplication in clones of *Aedes albopictus* cells, *J. Virol.*, 37, 889, 1981.
22. Sinarachatanant, P. and Olson, L. C., Replication of Dengue virus type 2 in *Aedes albopictus* cell culture, *J. Virol.*, 12, 275, 1973.
23. Kuno, G., Persistent infection of a nonvector mosquito cell line (TRA-171) with Dengue viruses, *Intervirology*, 18, 45, 1982.
24. Ng, M. L. and Westaway, E. G., Proteins specified by togaviruses in infected *Aedes albopictus* (Singh) mosquito cells, *J. Gen. Virol.*, 43, 91, 1979.
25. Riedel, B. and Brown, D. T., Role of extracellular virus in the maintenance of the persistent infection induced in *Aedes albopictus* (mosquito) cells by Sindbis virus, *J. Virol.*, 23, 554, 1977.
26. Reigel, F. and Koblet, H., Selective disappearance of two secreted host proteins in the course of Semliki Forest virus infection of *Aedes albopictus* cells, *J. Virol.*, 39, 321, 1981.
27. Eaton, B. T., Persistent togavirus infection of *Aedes albopictus* cells, in *Viruses and Environment*, Kurstak, E. and Maramorosch, K., Eds., Academic Press, San Francisco, 1978, 181.
28. Davey, M. W. and Dalgarno, L., Semliki Forest virus replication in cultured *Aedes albopictus* cells: studies on the establishment of persistence, *J. Gen. Virol.*, 24, 453, 1974.
29. Stalder, J., Reigel, F., and Koblet, H., Defective viral RNAs in *Aedes albopictus* C6/36 cells persistently infected with Semliki Forest virus, *Virology*, 129, 247, 1983.
30. Igarashi, A. and Stollar, V., Failure of defective interfering particles of Sindbis virus produced in BHK or chicken cells to affect viral replication in *Aedes albopictus* cells, *J. Virol.*, 19, 398, 1976.
31. Strauss, E. and Strauss, J. H., Mutants of alphaviruses: genetics and physiology, in *The Togaviruses, Biology, Structure, Replication*, Schlesinger, R. W., Ed., Academic Press, New York, 1980, 393.
32. Renz, D. and Brown, D. T., Characteristics of Sindbis virus temperature-sensitive mutants in cultures of BHK-21 and *Aedes albopictus* (mosquito) cells, *J. Virol.*, 19, 775, 1976.
33. Eaton, B. T., Heterologous interference in *Aedes albopictus* cells infected with alphaviruses, *J. Virol.*, 30, 45, 1979.
34. Enzmann, P.-J., Induction of an interferon-like substance in persistently infected *Aedes albopictus* cells, *Arch. Gesamte Virusforsch.*, 41, 382, 1973.
35. Murray, A. M. and Morahan, P. S., Studies of interferon production in *Aedes albopictus* mosquito cells, *Proc. Soc. Exp. Biol. Med.*, 142, 11, 1973.
36. Kascsak, R. J. and Lyons, M. J., Attempts to demonstrate the interferon defense mechanism in cultured mosquito cells, *Arch. Gesamte Virusforsch.*, 45, 149, 1974.
37. Stollar, V., Shenk, T. E., Koo, R., Igarashi, A., and Schlesinger, R. W., Observations on *Aedes albopictus* cell cultures persistently infected with Sindbis virus, *Ann. N.Y. Acad. Sci.*, 266, 214, 1975.
38. Riedel, B. and Brown, D. T., Novel antiviral activity found in the media of Sindbis virus-persistently infected mosquito (*Aedes albopictus*) cell cultures, *J. Virol.*, 29, 51, 1979.
39. Lee, C.-H. and Schloemer, R. H., Mosquito cells infected with Banzi virus secrete an antiviral activity which is of viral origin, *Virology*, 110, 402, 1981.

Chapter 17

FUSION OF ALPHAVIRUS-INFECTED MOSQUITO CELLS

H. Koblet, A. Omar, and C. Kempf

TABLE OF CONTENTS

I.   Introduction 78

II.   Materials and Methods 78

III. Results 79
    A.   Basic Findings 79
        1.   Cell Lines, Virus Production, and Spontaneous pH
            Changes of the Medium 79
        2.   Time Course of Viral Infection and Fusion 79
    B.   Temperature and Duration of Exposure to pH 6 80
        1.   Variation of Temperature at pH 6 80
        2.   Variation of Exposure Time to pH 6 at Constant
            Temperature (28°C) 80
        3.   Combinations of Variable Exposure Time to pH 6 and
            Variable Temperatures 80
        4.   Morphology and Duration of Fusion 80
    C.   Inhibitors of Fusion 80
    D.   Fusion from Within vs. Fusion from Without and Turnover
        of a Putative Fusogenic Factor 84
        1.   Inhibitors of Group A 84
        2.   Inhibitors of Group B 85
        3.   Final Arguments for a Fusion from Within 85
    E.   Dissection of the Fusion Process 85
        1.   Aggregation 85
        2.   Triggering..85
        3.   Protease Inhibitors and the Search for Additional
            Cleavages of Viral Proteins...86
        4.   The Conformational Change is Stable and Irreversible..........86

IV. Discussion 87

V.   Summary...88

Acknowledgments 88

References 88

I. INTRODUCTION

This chapter deals with low pH-induced fusion of *Aedes albopictus* cells infected with Semliki Forest virus (SFV). SFV is a well studied virus and many excellent reviews have appeared on the epidemiology, host range, structure of the virion, infection, replication, and cytopathic effects (CPE).[1-5]

Membrane fusion reactions are of fundamental importance in biology. They form the basis of steady-state unification and separation of cellular compartments, as in endocytosis and exocytosis, muscle cell formation, and fertilization.

Fusion between the bilayers of enveloped viruses and plasma membranes or endosomal membranes is the mechanism by which the viral genetic information is introduced into the cell.[6] In several cases there is a correlation between fusogenicity and infectivity.[7] The viral spike proteins play a crucial role in catalyzing this reaction. Much is known about fusogenic proteins, but the molecular events leading to membrane fusion are largely unknown.[6,8-10]

Formally, there are three stages of fusion: (1) membrane to membrane contact as in cellular aggregation, (2) coalescence involving fusion of hydrophobic interiors of two membranes, and (3) separation.[8] From a geometric point of view there are planar and annular fusions.[6] In planar fusion the membranes of two compartments combine. Viral fusions are planar fusions. There are two known types of viral-induced cell-cell fusion. "Fusion from without" is a series of fusions between free virions and cell membranes culminating in cell to cell fusion. It is an early event occurring at the time of infection (early fusion), needs high multiplicities of infection (MOI), and is not prevented by inhibitors of macromolecular synthesis. "Fusion from within" is a late phenomenon occurring after the eclipse at the time of viral maturation at the plasma membrane (late fusion), it is independent of the MOI, and is restricted by inhibitors of macromolecular synthesis. This process is again a series of fusions. Cytoplasmic vesicles containing viral membrane proteins with the N-terminal in the lumen fuse with the plasma membrane in such a way that the N-terminal is on the outer side; the plasma membranes of neighboring cells then fuse.[6,9,11-13] Fusion from within and fusion from without therefore do not define the mechanism of cell to cell fusion; they only distinguish the topographical origins of the fusogenic proteins in the plasma membrane. SFV infected *A. albopictus* cells provide an ideal system to study fusion. The cells can be grown at pH 6 and pH 7 in monolayers and in spinner culture, and survive the infection as well as the cell to cell fusion. Thus, the behavior of the proteins at the cell surface can be studied before, during, and after fusion, and in monolayers, as single cells and in spinner culture.

II. MATERIALS AND METHODS

The following cell lines were examined: *Aedes albopictus* by Singh (ATCC CL 126, Flow Laboratories, Inc., 62nd passage, "A-cells"); *A. albopictus* Singh, clone C6/36 Igarashi (15th passage, "B-cells");[14] *A. pseudoscutellaris*, mos 61[15] (58th passage, "C-cells"); and *A. albopictus* Singh[16] (41st passage, "D-cells"). B, C, and D were gifts from Dr. S. Buckley (Yale Arbovirus Research Unit, New Haven, Conn.). Cells were grown in Mitsuhashi-Maramorosch Medium (MM) at 28°C without bicarbonate/CO_2.[17] Phosphate buffered saline (PBS), pH 6 or 7, was adjusted with NaCl to 400 mOsm. Semliki Forest virus (Kumba) was a gift from Dr. J. de Boer, Plum Island, N.Y. After a viral passage in C6/36 cells, the medium was collected 20 hr after infection (p.i.) and stored at −80°C. The titers were 10^9 to 10^{10} PFU/mℓ. In all of the experiments, cells were infected with 10 to 100 PFU/cell. Fusion was assayed by replacing the medium with MM or PBS at pH 6 and containing 25 mM glucose. After 30 min at 28°C, fusion was examined in the light microscope. Since more than 90% of the cells fused, no attempt was made to evaluate fusion quantitatively.

III. RESULTS

A. Basic Findings

1. Cell Lines, Virus Production, and Spontaneous pH Changes of the Medium

A subline of A-cells was observed which formed a syncytium encompassing the whole monolayer 18 to 20 hr after SFV infection.[18] Parts of the syncytium had disappeared 3 weeks later and a chronically infected culture developed. The formation of syncytia was independent of the MOI and the virus source (*Aedes,* chick, mouse brain). The viruses generated in the course of acute infection produced uniform-sized plaques. The most important observation was that spontaneous syncytia occurred only if the cultures were infected before day 8 after seeding; later infection did not provoke fusion. At the time of fusion the pH of the culture medium was below 6.3. To investigate the phenomenon, several parameters in different *Aedes* lines were studied. Using a seeding density of 10^4 cells/cm^2, we obtained the following results. The growth rates of the *uninfected* lines determined during 14 days after seeding were A > D = B > C. The pH of the medium decreased gradually in all of the cell lines. The lowest pH was reached in A-cells, 7 to 10 days after seeding and after formation of a monolayer. The time required to reach the lowest pH was dependent on the seeding density. Afterwards the pH increased up to 8.5 following 25 days of culture. The lowest pH obtained depended on that of the medium at the time of seeding. The pH drop was stereotypic and was always about 0.7 pH units for A-cells. However, if cells were grown at pH 6, the pH increased. In *infected* cells the highest virus titers were always produced 20 to 24 hr p.i. in nearly confluent 5-day-old monolayers. At pH 7, B-cells generated 250, D-cells generated 120 to 140, C-cells 12 to 14, and A-cells 1 to 3 PFU/cell. No significant changes in cell growth rates were observed. When 5-day-old cultures were infected, pH changes were similar to those of uninfected cells. However, the ΔpH of the culture medium at the time of seeding and the lowest pH reached was 0.9 pH units, as compared to 0.7 units in uninfected A-cells. If the cultures were infected when the pH was low, syncytia were formed. Infection was an absolute prerequisite for fusion since uninfected cells never fused. The virus titer attained was unimportant. Changing the pH of the medium of infected cultures intentionally yielded the same results. The cells fused at a pH of 6.1 or below within 30 min. Infected C6/36 (B-)cells exhibited a CPE at pH 7 with a star-like appearance which is not a fusion.[14] They fused only at pH 6.1. In contrast, fusion was either discrete or lacking in C-cells, even at a low pH. In conclusion, the most important conditions for cell to cell fusion are SFV infection and low pH.

2. Time Course of Viral Infection and Fusion

The sequence of events leading to fusion was studied in C6/36 cells.[19] Monolayers were infected at pH 7 and the pH of the medium was then adjusted to 6 at hourly intervals. No fusion was observed before the 7th hr p.i. Discrete foci of fusions appeared at 7 hr p.i. From this time onward the number of fusion-competent cells increased until the 16th hr p.i., when more than 90% of the cells fused. This development was not disturbed by pH 6. If the pH was lowered at any time p.i. and incubation continued at pH 6, then more than 90% of the cells fused at 16 hr p.i. However, if the pH increased above 6.3 at any time between 8 and 16 hr p.i., then the fusion was arrested at a given level. At pH 6 infection was possible but no immediate fusion was observed. Expression of the fusogenic potential was delayed to 30 hr p.i. and virus was released at a reduced rate. This may be explained by a rapid inactivation of virions at low pH (initial PFU/mℓ it was 5×10^6; after 45 hr incubation in MM at pH 6.9 and 28°C it was 10^6; and at pH 6.1 and 28°C it was 10^4) and a reduced infection- or replication rate. At pH 7 the onset for expression of fusion correlated with the synthesis of viral products. At 3 hr p.i. the 49S and the 26S mRNA were detectable. Viral proteins were visible at 6 hr and very prominent in fluorograms at 8 hr p.i.; they gave a positive immunofluorescence at the plasma membrane in some cells around 7 hr p.i. After infection with a MOI of ten, the PFU/mℓ was 10^5 (7 hr p.i.), 2×10^6 (8 hr), 8×10^7 (10 hr), 10^8 (12 hr), and 10^9 (22 hr). These results so far demonstrate that at pH 7, expression of a fusogenic factor requires 7 to 16 hr and develops in a nonsynchronized fashion, and the time for completion of fusion as visualized in the light microscope takes 30 min after lowering the pH. These data suggest that viral proteins are involved in fusion and that this fusion is a late one.

B. Temperature and Duration of Exposure to pH 6

1. Variation of Temperature at pH 6

To find the limiting temperature of fusion, C6/36 monolayers were infected with SFV for 16 hr at 28°C. The medium was then replaced with MM of pH 6 at various temperatures and kept at these temperatures. Patchwise fusion was seen at 15°C; at 17°C complete syncytia still formed. Below 15°C fusion was never observed.[20]

2. Variation of Exposure Time to pH 6 at Constant Temperature (28°C)

It has been shown in other systems that after exposure to pH 6, fusion also proceeds at pH 7. The minimal time of exposure to pH 6 necessary to elicit fusion at pH 7 was therefore determined. *Aedes* cells grown on cover slips at 17 hr p.i. were dipped for various times into MM, pH 6, and replaced in MM, pH 7. After an exposure of 15 sec, fusion was complete after 30 min. Even 1 sec was sufficient, however, the fusion process was complete only after 2 hr. This delay probably reflects a nonuniform pH equilibration at the cell surface.[20]

3. Combinations of Variable Exposure Time to pH 6 and Variable Temperatures

Exposure times to pH 6 at temperatures below 15°C were varied up to 3 hr. Upon warming the cultures to 28°C, it always took 30 min for fusion to be complete. Furthermore, a few seconds at pH 6, even at 4°C, were sufficient to elicit the fusion at pH 6 or 7 and 28°C.[20]

Taken together, these results demonstrate two things, first that fusion is initiated by a triggering event which is rapid, pH dependent, and temperature independent, and it probably occurs at the outer surface of the plasma membrane. Secondly, below 15°C fusion is blocked at an early step after the triggering event; this cannot be overcome by prolonged incubation. This step is temperature dependent and pH independent and seems to be slow. The fusion process can be dissected into phases. This will be discussed in more detail in Section III.E.

4. Morphology and Duration of Fusion

The statement that the second step leading to syncytium formation is slow is correct only with respect to light microscopic examination. In fact, fusion is a fast process as perceived with the electron microscope.[21] Small intermembranous bridges become visible within 30 sec after lowering the pH (Figure 1).

C. Inhibitors of Fusion

All of the results described above indicate that fusogenic factors are produced and

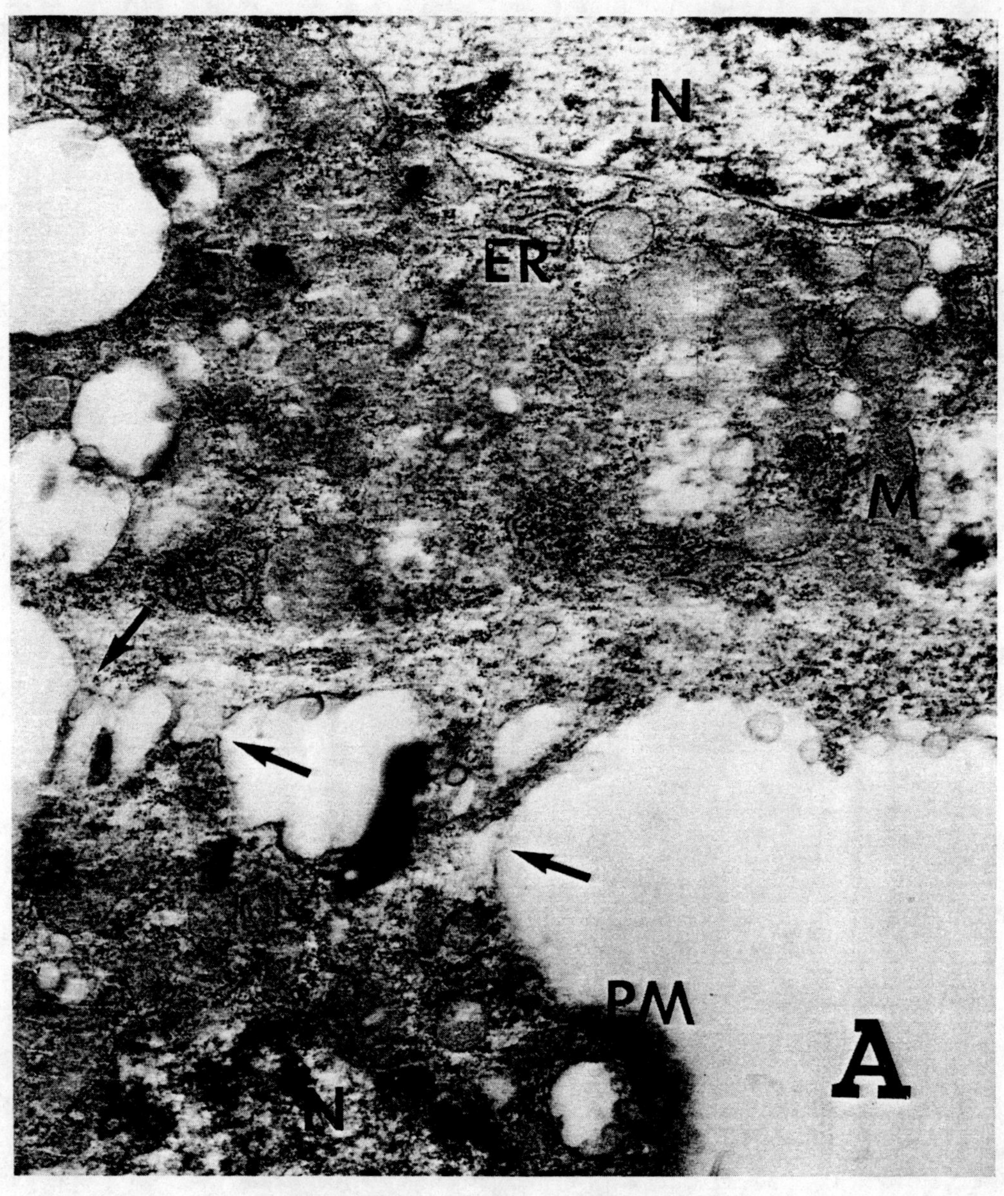

A

FIGURE 1. Confluent monolayers of C6/36 cells were infected with Semliki Forest virus (MOI = 20). At 16 hr p.i. the cells were exposed for 45 sec (A) and 2 min (B) to pH 6 and then immediately fixed with 2.5% glutaraldehyde at 4°C. Further processing of the monolayers for electron microscopy was performed according to established procedures. ER: endoplasmatic reticulum, M: mitochondria, N: nucleus, PM: plasma membrane, and V: SFV in intracellular vacuoles. Arrows indicate fusion bridges formed between neighboring cells (in A) and extended laterally (in B). (Magnification: × 14,842.)

integrated into the plasma membrane during the course of SFV replication. This was further investigated through the use of inhibitors. There is a rich body of literature describing inhibitors of fusion. These substances are grouped as follows: (1) inhibitors of virus uptake, replication, biosynthesis of macromolecules, and maturation (group A); (2) chemicals acting directly at or in the plasma membrane (group B); and (3) other

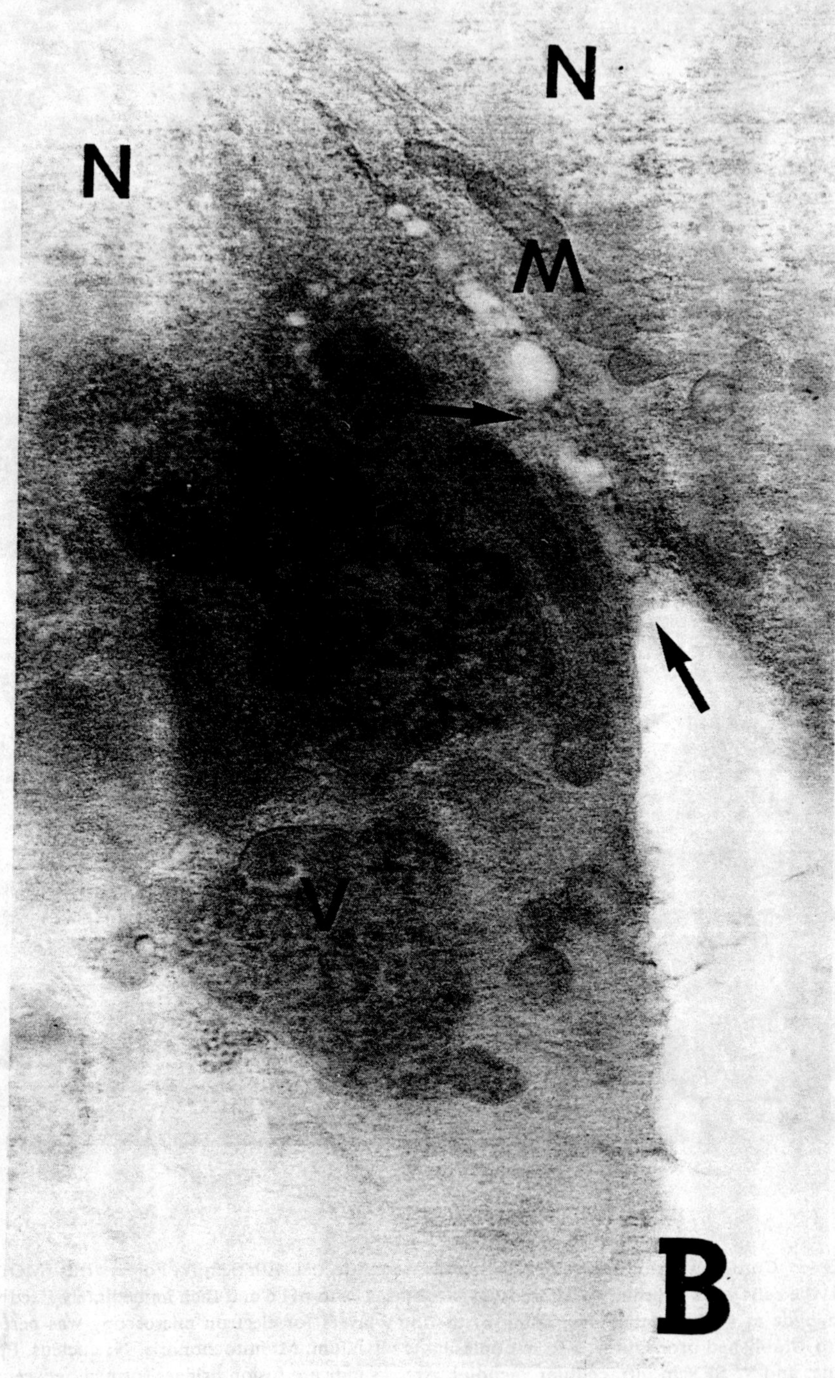

FIGURE 1B

inhibitors (energy metabolism, cytoskeletal). The most important inhibitors are given in Table 1. In our system, several potentially interesting substances proved to be ineffective as inhibitors. These were the amino acid analogues L-canavanine and p-fluorophenylalanine (5 mM), all drugs acting on the cytoskeleton (colcemide, cytochalasin B, podophyllotoxin, vinblastine, vincristine), some proteolytic enzymes (trypsin, chymo-

Table 1
CHEMICALS USED AS POSSIBLE INHIBITORS OF FUSION (C6/36) CELLS

Name[a]	Concentration used	Reversibility[b]	Ref.
Group A			
Lysosomotropic agents (inhibitors of uncoating)			
Chloroquine[c]	20 mM	Yes	19, 24, 26
Inhibitors of protein biosynthesis			
Cycloheximide[d]	50 µg/ml	Yes	11, 19, 26
Inhibitors of protein glycosylation			
Tunicamycin[e]	0.5 µg/ml	Yes	9, 19, 26
Group B			
Ionophores[f]			
A 23 187	10 µM	No	26
Dianemycin	10 µM	Yes	26
Monensin	10 µM	Yes	19, 26
Nigericin	10 µM	Yes	26
Valinomycin	10 µM	No	26
CCCP	10 µM	No	26
TCSA	10 µM	Yes	26
EGTA	1 mM	Yes	22, 26
Anesthetics[f]			
Chlorpromazine	100 µM	n.d.	52, 53
Dibucaine	100 µM	n.d.	52
Proteolytic enzymes			
Bromelaine[f]	200 µg/ml		26
Protease inhibitors and analogs			
TPCK[f]	1—5 mM	No	19, 22, 26, 53
Protein modifying agents[g]			
Tetrathionate	1—10 mM	Yes	20, 22, 26
DTNB	20 mM	No	
DTT	1—10 mM	Yes	26
TNBS	5 mM	No	
Butandione	1—10 mM	No	26
Cyclohexandione	10 mM		
Phenylglyoxal	10 mM		
Group C			
Energy metabolism, proton pump[f]			
Sodium azide	≥1 µM	n.d.	26

[a]   CCCP is carbonylcyanide-m-chlorophenylhydrazone; TCSA is 3,3′,4′,5-tetra-chlorosalicylanilide; EGTA is ethyleneglycolbis (α-aminoethylether) N,N,N′,N′-tetraacetic acid; TPCK is N-α-tosyl-L-phenylalanyl-chloromethylketone; DTNB is 5,5′-dithiobis (nitrobenzoic acid) (Ellman); DTT is dithiothreitol; and TNBS is trinitrobenzenesulfonic acid.

[b]   n.d. = not done.

[c]   Fusion is inhibited when added at 0 to 1 hr p.i.

[d]   Fusion is inhibited when added at 0 to 6 hr p.i.

[e]   Fusion is inhibited when added at 0 to 3 hr p.i. (partial).

[f]   Fusion is inhibited when added at all times p.i.

[g]   Fusion is inhibited when added at all times, but requiring a pH exposure.

trypsin, thermolysin, proteinase K), most protease inhibitors (p-tosyl-L-arginine-methylester, N-α-tosyl-L-lysyl-chloromethylketone, leupeptin, phenylmethylsulfonylfluoride),[22] and a competitive inhibitor of fusion, carbobenzoxy-D-phenylalanyl-L-phenylalanyl-nitro-L-arginine [SV 4814]).[23] The lysosomotropic agent NH_4Cl (30 mM) neither prevented virus formation nor inhibited fusion.[24] Puromycin (100 μg/mℓ) inhibited protein synthesis only to 70% in C6/36 cells which proved impractical.

D. Fusion from Within vs. Fusion from Without and Turnover of a Putative Fusogenic Factor

The appearance of viral products correlated with the onset of the capacity for fusion, and this in turn is indicative of a fusion from within. The inhibitors of fusion yield additional evidence for this thesis.

1. Inhibitors of Group A

In vertebrate cells lysosomotropic agents inhibit the fusion of the envelope of the endocytosed virions with the endosomal membrane (the uncoating) by alkalization of the content of the vesicles.[24] Presumably the mechanism is the same in invertebrate cells.[25] If chloroquine was added at the time of infection, a direct dependence of the virus titer on the concentration of the agent was obtained. At 10 mM the viral titer was 30 PFU/mℓ. These cells still fused at pH 6. Only at 20 mM was fusion inhibited. If chloroquine was removed at 16 hr p.i., then the cells fused 15 hr later (31 hr p.i.) at pH 6. From these experiments we concluded that the fusion is compatible with a fusion from within, C6/36 cells are rather resistant to lysosomotropic agents, fusion is possible at very low rates of virus production, and the action of chloroquine is reversible; virions seem to remain in the cells in a "frozen state".[19] In other experiments, C6/36 cells were infected and cycloheximide (Cy) was added at various times at pH 7.[19,26] At 16 hr p.i. the medium was replaced with MM, pH 6, without Cy. Fusion was complete after 30 min only when Cy was present from 8 hr p.i. onwards. This result shows that protein biosynthesis must be undisturbed for 7 to 8 hr p.i. in order to produce the fusogenic factor; this substantiates a fusion from within. A mere toxic effect of Cy was ruled out by adding the drug from the time of infection to 16 hr p.i. and then removing it. Fusion was observed 8 hr later at 24 hr p.i. Conclusively, Cy does not prevent uptake of virions, the action of Cy is reversible, Cy inhibits SFV replication at an early stage without damaging the genetic information because the delay between lifting the inhibition and the onset of fusion corresponds to the normal time course of replication, and the fusogenic factor(s) remains potent for a long time once it is synthesized. Similar findings have been reported for Newcastle disease virus.[9,11] However, it is conceivable that fusogenic factors should be exhaustible if their renewal is blocked by Cy. Therefore, after appearance of the capacity for fusion, cells were labeled with short pulses of ^{35}S-methionine at pH 7. Chases were performed with and without Cy for up to 10 hr at pH 7. At the end of the chase, fusion assays or polyacrylamide gel electrophoresis (PAGE) were done. The following results were obtained: the viral products are visible in the fluorograms even after a 10 hr chase with Cy and only decrease thereafter, without Cy the decrease sets in earlier at 3 to 4 hr chase, and fusion is greatly reduced in the presence of at least 8 hr of Cy.

All known fusogenic factors in enveloped viruses have been reported to be glycoproteins.[9] Furthermore, the oligosaccharide moiety seems to play a crucial role in the fusion process. Therefore it was important to know whether the same holds true in our system. Tunicamycin (Tm) is an inhibitor of protein glycosylation in vertebrates and invertebrates.[27] Tm was added at various times p.i. and fusion assays were performed 24 hr p.i.[19,26] Fusion was reduced, but still detectable, if Tm was added between 0 and 3 hr p.i. Fusion was never totally abolished. An escape glycosylation of viral proteins was improbable since PAGE showed that Tm prevented the glycosylation of SFV membrane proteins as well as the cleavage of the p 62 protein. Apparently glycosylation (in *Aedes)* and cleavage of p 62 are unimportant for fusion. The action of Tm was fully reversible. If it was added from 0 to 24 hr p.i. and washed away with MM pH 6, complete syncytia appeared at 31 hr p.i.

2. Inhibitors of Group B

Ionophores permit free permeability of the respective ions they carry[28] resulting in a dissipation of existing ionic gradients across the membrane. All of the ionophores tested prevented fusion immediately if added concomitantly in MM, pH 6, at any time p.i.[19,26] The speed of action favors the assumption that the ionophores act directly at the plasma membrane nonspecifically. Identical fluorograms were obtained after 90 min of monensin treatment as without treatment. Monensin also furnished excellent evidence for a fusion from within. One could argue that the interpretation of previous experiments is wrong because at 8 hr p.i. sufficient virions are shed to promote a fusion from without, and monensin does not act on a fusogenic factor in the cell membrane, but prevents binding of exogenous virions which had been shed in the course of infection. The adsorption of labeled virions to infected and uninfected C6/36 cells at pH 6 and 7 with and without monensin was measured. The results showed that infected cells bind fewer labeled virions than uninfected cells at pH 7 and 4°C; however, in the presence of monensin, virus binding at pH 6 is restored to the niveau of uninfected cells at pH 7, yet there is no fusion.[19]

Proteolytic enzymes were used to digest the N-terminals of the viral spikes on the outside of the plasma membrane. Only bromelain inhibited fusion to a large extent.

3. Final Arguments for a Fusion from Within

The simplest way to assay for a fusion from without is to add virions to a cell culture.[11] Up to 1000 PFU/cell ($>10^4$ particles) were applied to uninfected cells for 1 hr at 4°C (binding without uptake); then MM, pH 6 at 28°C was added. Only a partial fusion was obtained with a MOI of 1000 and more.[19]

Furthermore, if labeled virions were added to uninfected C6/36 cells for 1 hr at 4°C and then digested in PBS with proteinase K (200 to 500 $\mu g/m\ell$) at 4°C for 1 hr,[24] virtually all of the label was removed. When the same procedure was performed on infected cells at 16 hr p.i. and then MM, pH 6 28°C was added, syncytia formed rapidly.[19] Proteinase K digests exogenous virions bound to plasma membranes, but does not destroy fusion factors in the membrane. Thus, we conclude that fusion is possible in the absence of exogenous virus, confirming a fusion from within.

E. Dissection of the Fusion Process

1. Aggregation

Infecting a spinner culture containing about 4×10^5 cells/mℓ with a MOI of 100 in MM, pH 7 caused the cells to aggregate at about 4 hr p.i. If the pH of the medium was lowered to 6 at hourly intervals p.i., the cells began to fuse as in the monolayers at 7 hr p.i. Spinner cultures, therefore, offer a convenient means for studying the two individual processes of aggregation and fusion independently. Moreover, it shows that fusion follows an aggregation of the cells.

2. Triggering

In Section B.3, it was shown that pH 6 serves to trigger the fusion process.[20] Some of the inhibitors in Table 1 are perfect tools to study the trigger mechanism in more detail. As mentioned above, monensin inhibits fusion immediately at any time. The fact that after removal of monensin syncytia developed at pH 7 implied that its action takes place after the triggering event. This is compatible with the assumption of the earliest step being a conformational change of a cell surface structure — most probably a viral-induced protein — as described for Sindbis virions.[29-31] Therefore, chemical modifications of surface proteins were investigated (Table 1, group B, protein modifying agents).

Tetrathionate reacts covalently with sulfhydryl groups. If it was added to cells 16 hr p.i. at pH 7, 28°C and then the unreacted reagent removed, fusion occurred unimpaired after lowering the pH to 6. However, if the reagent was present at pH 6, fusion was prevented. Therefore, it is conceivable that at pH 6 additional SH-groups become exposed and available for modification, the blockage of which prevents fusion. With this consideration, infected cells were exposed to pH 6 at 4°C where no fusion occurs, and then incubated with tetrathionate at pH 7 and 28°C. No fusion was observed. Thus, a previous exposition to pH 6 is mandatory for the exposition of additional SH-groups. Furthermore, if the tetrathionate derivative formed at pH 6 was cleaved with DTT and then both reagents removed, fusion developed at pH 6 and 7.[20] Ellman's reagent reacted in a similar way as tetrathionate. Trinitrobenzenesulfonic acid (TNBS) and butandione had the same effect as tetrathionate in preventing fusion only if the reaction was carried out after the pH change. All of these experiments show that the pH change is accompanied by a conformational change and the exposure of thiol-, amino-, and guanidino-groups on the cell surface which are essential for fusion.

3. Protease Inhibitors and the Search for Additional Cleavages of Viral Proteins

Though SFV-membrane proteins must be intact to be fusogenic in the virion,[32] an unprecedented cleavage of a spike protein similar to that in myxo- or paramyxoviruses[6,7,10,33] was sought on the basis that N-α-tosyl-L-phenylalanyl-chloro-methylketone (TPCK) inhibited fusion in a rapid and irreversible manner (30 min, pH 7).[22] Pulse-chase experiments showed no difference in the banding pattern in PAGE of the controls and TPCK-treated cultures. Protein 62 was properly cleaved in the presence of TPCK 30 min after the start of the pulse.[19] Pulse-chase experiments without TPCK at pH 6 or 7 produced the same results in PAGE even if cell extracts were alkylated or reduced and alkylated.

4. The Conformational Change is Stable and Irreversible

Several conformational states may be possible in the case of transmembrane proteins. A final conformation or a transitional state could trigger fusion. The pH induced conformation change could be reversible. To answer this, C6/36 monolayers were infected and at 16 hr p.i. the pH was lowered to 6. Then uninfected cells vitally stained with 1% toluidine blue were added at various times up to 240 min later. At all times the stained cells fused with the monolayer or the preformed syncytia, resulting in the distribution of the dye. The same result was found if protein synthesis was blocked in both partners with cycloheximide 30 min before lowering the pH to prevent the continuous arrival of newly formed proteins at the cell surface. These results prove that it is the final conformation which is fusogenic and stable.

To gain insight whether the new conformation is irreversible, SFV infected cells were exposed to pH 6 at 14°C for 10 min. The pH was then raised to 7 at 14°C. The cells were kept under these conditions for up to 2 hr to allow a possible rearrangement. At 14°C no fusion occurred; however as soon as the temperature was elevated to 28°C, syncytia were formed.[20] This strongly suggests that the conformational change is irreversible. In addition, all of these experiments explain the fact earlier described, that fusion proceeds undisturbed if the pH is lowered already during the eclipse.

IV. DISCUSSION

The minimal requirements and the biophysical theory of membrane fusion have been dealt with in several papers.[6,8,9,32-36] Therefore we will restrict ourselves to data pertinent to this treatise. We describe herein low pH-induced fusion studied for the first time in *Aedes albopictus* cells after infection with SFV. The cells are ideal for investigating the mechanism involved in cell membrane fusion. However, one should consider that several factors in these cell cultures need careful control because the pH threshold to trigger fusion is quite sharp. Otherwise it is unpredictable whether spontaneous fusion will develop. Recognized factors influencing pH, pH changes, and fusion are the cell line per se, the seeding density, the time of infection relative to the time of seeding, and the initial pH of the medium.[18,37]

Many results presented in this chapter provide strong evidence for a fusion from within. Though fusion from without can be enforced with extremely high MOIs, the digestion of exogenous virions in infected cultures with proteinase K did not inhibit fusion. Even restoration of the binding of exogenous virions to infected cell membranes by monensin did not lead to cell fusion. Fusion capacity develops only after the eclipse at the time when viral proteins appear at the cell surface. Cycloheximide inhibits this process if added soon after infection.

It has been shown that certain viruses hemagglutinate erythrocytes at low pH. In fact, temperature and pH conditions were used to distinguish a group A from a group B of arboviruses.[38] Hemagglutination (0 to 40°C) and hemolysis and fusion (37 to 42°C) of various erythrocytes by SFV and Sindbis virus were shown to proceed at pH 5.8.[39,40] In recent years a wide variety of systems was studied with SFV, Sindbis, vesicular stomatitis virus, Sendai, fowl plague, liposomes, and different cell lines.[6,32,41] From these reports it has become apparent that low pH plays a very significant role in these fusion processes. By utilizing different temperatures it is possible to dissect the fusion process in *Aedes* cells into two steps. The first step at the cell membrane leading to fusion is a conformational change of a surface structure which is pH-dependent and temperature independent. There are two ways of allowing this conformational change without permitting fusion: (1) keep the cells below 15°C, and (2) add inhibitors such as certain ionophores and protein-modifying agents. All of these inhibitors act at a step of the fusion chain following the conformational change. A conformational change is the more probable, as no difference in the protein pattern in fluorograms could be found. The conformational change is an irreversible activation of the fusion process. Similar conformational changes were reported for the Sindbis virion and for several other nonToga viruses.[29-31,34,42] However, this mechanism does not necessarily apply to all viral systems.[43,44]

The identity of the fusogenic factor is uncertain. The results with bromelaine and with the modifying compounds strongly suggest that it is a protein. Viral infection is mandatory for fusion and this favors the assumption of the involvement of a viral spike protein. In the Sindbis *virion,* low pH was found to induce a conformational change in the E_2 polypeptide.[29] However, even if E_2 cooperates in a spike,[45] our results and those of other authors favor the assumption that the fusogenic protein in alphaviruses is probably E_1. The reasons are manyfold: (1) the precursor of E_2, p62, may remain uncleaved;[19,46] (2) E_1 is the hemagglutinin[6,47] and also contains the hemolytic activity and the former determinant is located near the apex of the spike and the latter near the virion membrane;[48] (3) the external domains of E_1 in Sindbis and SFV possess a conserved hydrophobic sequence;[2] and (4) fusion is not possible if E_1 is not expressed on the cell surface.[12] Therefore the weight of the evidence points to E_1. The consensus is that fusogenic proteins are glycoproteins,[9] but exceptions have been described.[49] Our finding that tunicamycin does not inhibit fusion in *Aedes* cells while biochemically acting as expected may represent a special case of this system.[27]

Once fusion is triggered, it proceeds rapidly as seen with electron microscopy (Figure 1).[21] After a few minutes fusion bridges are observed. We assume that such microfusions are the prerequisite to overcoming the repulsive forces in the immediate neighborhood. The bridges then enlarge on all sides and it takes about 30 min until the fusion of large areas is visible in the light microscope. In view of the rapidity of the process, especially the triggering event, the question arises how the intracellular maturation of togavirions in *Aedes* cells (which is well documented)[50] can be integrated into this picture. In C6/36 cells, we have evidence for intracellular as well as plasma membrane viral maturation.[51] It is worthwhile noting how many common facets have been found in this interesting phenomenon between organisms widely separated in phylogeny.

V. SUMMARY

This chapter deals with low pH-induced fusion of *A. albopictus* cells infected with Semliki Forest virus. In the first part, the behavior of different *Aedes* cell lines in culture is described. It was observed that a gradual decrease in the pH of the medium during cultivation effected certain cell lines to fuse after infection. This phenomenon was influenced by the seeding density and the time of infection.

In the second part, results were presented showing that the fusion under investigation is a fusion from within. This is based on establishing a correlation between the time of appearance of viral products after the eclipse and the onset of fusion capacity. Inhibitors of macromolecular biosynthesis and drugs acting at the cell membrane were applied to verify this finding.

The final part treats the fusion per se. Using different methods, the fusion process could be dissected into two steps. The initial event is a conformational change triggered by low pH leading to an irreversible activation of a fusogenic state. Several arguments favor the assumption that the fusogenic protein in alphaviruses is the E_1 spike protein.

ACKNOWLEDGMENTS

This paper was supported by the Swiss National Science Foundation Grants No. 3.384.082 and No. 3.494.079.

We wish to thank Dr. P. Späth, Dr. F. Reigel, and Dr. A. Flaviano for their collaboration, Mrs. E. Boerlin for the electron microscopic work, Miss U. Kohler for skillful technical assistance, and Mrs. C. Lam for typing the manuscript.

REFERENCES

1. Schlesinger, R. W., *The Togaviruses, Biology, Structure, Replication*, Schlesinger, R. W., Ed., Academic Press, New York, 1980.
2. Garoff, H., Kondor-Koch, C., and Riedel, H., Structure and assembly of alpha viruses, *Curr. Top. Microbiol. Immunol.*, 99, 1, 1982.
3. Strauss, E. G. and Strauss, J. H., Replication strategies of the single stranded RNA viruses of eukaryotes, *Curr. Top. Microbiol. Immunol.*, 105, 1, 1983.
4. Shatkin, A. J., Molecular mechanism of virus-mediated cytopathology, *Philos. Trans. R. Soc. London, Ser. B*, 303, 167, 1983.
5. Yunker, C. E., Arthropod tissue culture in the study of arboviruses and rickettsiae: a review, *Curr. Top. Microbiol. Immunol.*, 55, 113, 1971.

6. White, J., Kielian, M., and Helenius, A., Membrane fusion proteins of enveloped animal viruses, *Q. Rev. Biophys.*, 16, 151, 1983.

7. Rott, R., Determinants of influenza virus pathogenicity, *Hoppe-Seyler's Z. Physiol. Chem.*, 363, 1273, 1982.

8. Knutton, S. and Pasternak, C. A., The mechanism of cell-cell fusion, *Trends Biochem. Sci.*, 4, 220, 1979.

9. Gallaher, W. R., Levitan, D. B., Kirwin, K. S., and Blough, H. A., Molecular and biological parameters of membrane fusion, in *Cell Membranes and Viral Envelopes*, Vol. 1, Blough, H. A. and Tiffany, J. M., Eds., Academic Press, London, 1980, 395.

10. Asano, A. and Asano, K., Mechanism of HVJ induced cell fusion, in *Structure, Dynamics and Biogenesis of Biomembranes*, Sato, R. and Ohnishi, S.-I., Eds., Japan Scientific Societies Press, Tokyo, and Plenum Press, New York-London, 1982, 57.

11. Bratt, M. A. and Gallaher, W. R., Preliminary analysis of the requirements for fusion from within and fusion from without by Newcastle disease virus, *Proc. Natl. Acad. Sci. U.S.A.*, 64, 536, 1969.

12. Kondor-Koch, C., Burke, B., and Garoff, H., Expression of Semliki Forest virus proteins from cloned complementary DNA. I. The fusion activity of the spike glycoprotein, *J. Cell. Biol.*, 97, 644, 1983.

13. Saraste, J. and Hedman, K., Intracellular vesicles involved in the transport of Semliki Forest virus membrane proteins to the cell surface, *EMBO J.*, 2, 2001, 1983.

14. Igarashi, A., Isolation of a Singh's *Aedes albopictus* cell clone sensitive to dengue and chikungunya viruses, *J. Gen. Virol.*, 40, 531, 1978.

15. Varma, M. G. R., Pudney, M., and Leake, C. J., Cell lines from larvae of *Aedes (Stegomyia) malayensis* (Colles) and *Aedes (s) pseudoscutellaris* (Theobald) and their infection with some arboviruses, *Trans. R. Soc. Trop. Med. Hyg.*, 68, 374, 1974.

16. Singh, K. R. P., Cell cultures derived from larvae of *Aedes albopictus* (Skuse) and *Aedes aegypti* (L.), *Curr. Sci.*, 36, 506, 1967.

17. Mitsuhashi, J., Media for insect cell cultures, *Adv. Cell Cult.*, 2, 133, 1982.

18. Späth, P. J. and Koblet, H., Alphavirus induced syncytium formation in *Aedes albopictus* cell cultures, in *Invertebrate Systems In Vitro*, Kurstak, E., Maramorosch, K., and Dübendorfer, A., Eds., Elsevier/North-Holland, Amsterdam, 1980, 375.

19. Omar, A., Flaviano, A., Kohler, U., and Koblet, H., Fusion of Semliki Forest Virus infected *Aedes albopictus* cells at low pH is a fusion from within, *Arch. Virol.*, 89, 145, 1986.

20. Koblet, H., Kempf, C., Kohler, U., and Omar, A., Conformational changes at pH 6 on the cell surface of Semliki Forest virus infected *Aedes albopictus* cells, *Virology*, 143, 334, 1985.

21. Knutton, S., Studies of membrane fusion. VI. Mechanism of the membrane fusion and cell swelling stages of Sendai virus-mediated cell fusion, *J. Cell. Sci.*, 43, 103, 1980.

22. Ahkong, Q. F., Botham, G. M., Woodward, A. W., and Lucy, J. A., Calcium-activated thiol-proteinase activity in the fusion of rat erythrocytes induced by benzyl alcohol, *Biochem. J.*, 192, 829, 1980.

23. Richardson, C. D., Scheid, A., and Choppin, P. W., Specific inhibition of Paramyxovirus and Myxovirus replication by oligopeptides with amino acid sequences similar to those at the N-termini of the F_1 or HA_2 viral polypeptides, *Virology*, 105, 205, 1980.

24. Helenius, A., Marsh, M., and White, J., Inhibition of Semliki Forest virus penetration by lysosomotropic weak bases, *J. Gen. Virol.*, 58, 47, 1982.

25. Coombs, K., Mann, E., Edwards, J., and Brown, D. T., Effects of chloroquine and cytochalasin B on the infection of cells by Sindbis virus and vesicular stomatitis virus, *J. Virol.*, 37, 1060, 1981.

26. Flaviano, A., Zell-Zell Fusion von Semliki Forest Virus Infizierten *Aedes albopictus* Zellen, Ph.D. thesis, University of Berne, Switzerland, 1984.

27. Hsieh, P. and Robbins, P. W., Regulation of asparagine-linked oligosaccharide processing. Oligosaccharide processing in *Aedes albopictus* mosquito cells, *J. Biol. Chem.*, 259, 2375, 1984.

28. Pressman, B. C., Biological applications of ionophores, *Ann. Rev. Biochem.*, 45, 501, 1976.

29. Edwards, J., Mann, E., and Brown, D. T., Conformational changes in Sindbis virus envelope proteins accompanying exposure to low pH, *J. Virol.*, 45, 1090, 1983.

30. Kaluza, G., Rott, R., and Schwarz, R. T., Carbohydrate-induced conformational changes of Semliki Forest virus glycoproteins determine antigenicity, *Virology*, 102, 286, 1980.

31. Wolcott, J. A., Wust, C. J., and Brown, A., Identification of immunologically cross-reactive proteins of Sindbis virus: evidence for unique conformation of E1 glycoprotein from infected cells, *J. Virol.*, 49, 379, 1984.

32. White, J. and Helenius, A., pH-dependent fusion between the Semliki Forest virus membrane and liposomes, *Proc. Natl. Acad. Sci. U.S.A.*, 77, 3273, 1980.

33. Huang, R. T. C., Wahn, K., Klenk, H.-D., and Rott, R., Fusion between cell membrane and liposomes containing the glycoproteins of influenza virus, *Virology*, 104, 294, 1980.

34. Hsu, M.-C., Scheid, A., and Choppin, P. W., Fusion of Sendai virus with liposomes: dependence on the viral fusion protein (F) and the lipid composition of liposomes, *Virology,* 126, 361, 1983.

35. Connor, J., Yatvin, M. B., and Huang, L., pH-sensitive liposomes: acid-induced liposome fusion, *Proc. Natl. Acad. Sci. U.S.A.,* 81, 1715, 1984.

36. Blumenthal, R., Henkart, M., and Steer, C. J., Clathrin-induced pH-dependent fusion of phosphatidyl cholin vesicles, *J. Biol. Chem.,* 258, 3409, 1983.

37. Pudney, M., Leake, C. J., and Buckley, S. M., Replication of arboviruses in arthropod *in vitro* systems: an overview, in *Invertebrate Cell Culture Applications,* Maramorosch, K. and Mitsuhashi, J., Eds., Academic Press, London, 1982, 159.

38. Casals, J. and Brown, L. V., Hemagglutination with arthropod-borne viruses, *J. Exp. Med.,* 99, 429, 1954.

39. Väänänen, P. and Kääriäinen, L., Fusion and haemolysis of erythrocytes caused by three togaviruses: Semliki Forest, Sindbis and rubella, *J. Gen. Virol.,* 46, 467, 1980.

40. Lenard, J. and Miller, D. K., pH-dependent hemolysis by influenza, Semliki Forest virus, and Sendai virus, *Virology,* 110, 479, 1981.

41. White, J., Matlin, K., and Helenius, A., Cell fusion by Semliki Forest, influenza, and vesicular stomatitis viruses, *J. Cell. Biol.,* 89, 674, 1981.

42. Daniels, R. S., Douglas, A. R., Skehel, J. J., and Wiley, D. C., Analyses of the antigenicity of influenza haemagglutinin at the pH optimum for virus-mediated membrane fusion, *J. Gen. Virol.,* 64, 1657, 1983.

43. Sato, S. B., Kawasaki, K., and Ohnishi, S.-I., Hemolytic activity of influenza virus hemagglutinin glycoproteins activated in mildly acidic environments, *Proc. Natl. Acad. Sci. U.S.A.,* 80, 3153, 1983.

44. Yewdell, J. W., Gerhard, W., and Bächi, T., Monoclonal antihemagglutinin antibodies detect irreversible antigenic alterations that coincide with the acid activation of influenza virus A/PR/834-mediated hemolysis, *J. Virol.,* 48, 239, 1983.

45. Ziemiecki, A., Garoff, H., and Simons, K., Formation of the Semliki Forest virus membrane glycoprotein complexes in the infected cell, *J. Gen. Virol.,* 50, 111, 1980.

46. Mann, E., Edwards, J., and Brown, D. T., Polycaryocyte formation mediated by Sindbis virus glycoproteins, *J. Virol.,* 45, 1083, 1983.

47. Simizu, B., Yamamoto, K., Hashimoto, K., and Ogata, T., Structural proteins of chikungunya virus, *J. Virol.,* 51, 254, 1984.

48. Chanas, A. C., Ellis, D. S., Stamford, S., and Gould, E. A., The interaction of monoclonal antibodies directed against envelope glycoprotein E₁ of Sindbis virus with virus-infected cells, *Antiviral Res.,* 2, 191, 1982.

49. Wilcox, G. E. and Compans, R. W., Cell fusion induced by Nelson Bay virus, *Virology,* 123, 312, 1982.

50. Lehane, M. J. and Leake, C. J., A kinetic and ultrastructural comparison of Alphavirus infection of cultured mosquito and vertebrate cells, *J. Trop. Med. Hyg.,* 85, 229, 1982.

51. Simizu, B. and Maeda, S., Growth patterns of temperature-sensitive mutants of western equine encephalitis virus in cultured *Aedes albopictus* (mosquito) cells, *J. Gen. Virol.,* 56, 349, 1981.

52. Poste, G. and Reeve, P., Inhibition of virus-induced cell fusion by local anaesthetics and phenothiazine tranquillizers, *J. Gen. Virol.,* 16, 21, 1972.

53. Lucy, J. A., Fusogenic mechanisms, in *Cell Fusion,* Vol. 103, Ciba Foundation Symposium, Pitman, London, 1984, 28.

Chapter 18

ARBOVIRUS-VECTOR CELL INTERACTIONS AND HOST-RANGE VIRAL MUTANTS

V. Stollar

TABLE OF CONTENTS

I. Introduction 92

II. Description of Mosquito Cells Used for the Study of Arboviruses 92

III. Outcome of Infection of *Aedes albopictus* Cells with Different
       Viruses 93
       A. Family: Togaviridae 93
             1. Genus: *Alphavirus* 93
             2. Genus: *Pestivirus* 94
       B. Family: Flaviviridae 94
       C. Family: Bunyaviridae 95
       D. Family: Rhabdoviridae 95

IV. Biochemical Correlates of Cell Killing in VSV-Infected
       Aedes albopictus Cells 95

V. Host-Range Mutants of Sindbis Virus 102
       A. Mutants Restricted in Mosquito Cells 102
       B. A Mutant of Sindbis Virus Able to Replicate in Mosquito
             but not Vertebrate Cells 104

VI. Conclusion 107

Acknowledgments 108

References 108

I. INTRODUCTION

It is now close to 20 years since the report of Singh[1] in which he described the establishment of a cell line from larvae of the mosquito *Aedes albopictus*. Very soon thereafter in his laboratory and elsewhere, it was shown that various arthropod-borne arboviruses were able to replicate in mosquito cell lines, giving yields comparable to those obtained with vertebrate cells.[2,3]

Of the approximately 360 different arboviruses which have been identified, the great majority (280) can be placed into five different families: Togaviridae, Flaviviridae, Bunyaviridae, Rhabdoviridae, and Reoviridae.[4] An earlier review[3] summarized work which had been done with togaviruses in cultured arthropod cells. Although much of what will be covered in this review will deal with togaviruses, reference will also be made to recent work with viruses of the other families.

It is worth asking at this point why in addition to studying arboviruses in various vertebrate cell types, one might also wish to study these viruses in arthropod cells derived from known vectors. First, since the biochemical composition (specifically carbohydrates and lipids) of these viruses differs remarkably depending on whether they are grown in a vertebrate or arthropod host,[5-7] we would like to know whether these differences influence the biological or antigenic properties of the virus. Second, the observation that alphaviruses, which are generally quite cytopathic for vertebrate cells, can grow to high levels in mosquito cells without any cytopathic effect (CPE)[2] suggested that comparative studies of viral replication in these different cell types might shed light on the mechanisms of cell killing by these viruses. Third, studies on viral replication in cultured mosquito cells might be instructive as to the features of viral replication in the intact adult mosquito, and lastly, the ability to grow arboviruses in cells which are so widely separated phylogenetically, as mosquito and mammalian cells, offers an excellent opportunity to select for host-range viral mutants.

In this chapter I shall not attempt to present an exhaustive review listing every system which has been described. The reader is referred to earlier reviews[3,8-10] and to various other chapters in this book. Instead, my goals will be (1) to concentrate on representative systems which have been studied in some detail, so as to illustrate the different types of virus-cell interactions that have been observed in mosquito cells, and (2) to describe viral mutants which have a restricted host range and are unable to replicate efficiently either in arthropod cells or in vertebrate cells.

Since the replication of viruses in tick cell lines will be covered in another chapter, this discussion will be limited to viruses which have been studied in mosquito cells.

II. DESCRIPTION OF MOSQUITO CELLS USED FOR THE STUDY OF ARBOVIRUSES

Arboviruses have been grown in many different mosquito cell lines. However, at the present time most workers use cells derived from the *Aedes albopictus* line of Singh. Most of this discussion therefore will deal with experiments done with these cells.

Singh established his *A. albopictus* line from cells which had been prepared from freshly hatched larvae.[1] The medium used to grow these cells (MM medium)[11] was undefined, with the essential amino acids and vitamins being supplied in the form of lactalbumin hydrolysate and yeastolate. Since the cells grew very well in this medium, workers in many other laboratories continued to maintain these cells in MM medium. Later, however, these cells were adapted to Eagle's minimal medium supplemented with nonessential amino acids;[12] both the MM medium and the Eagle's medium were supplemented with fetal calf serum. Following the adaptation to Eagle's medium the

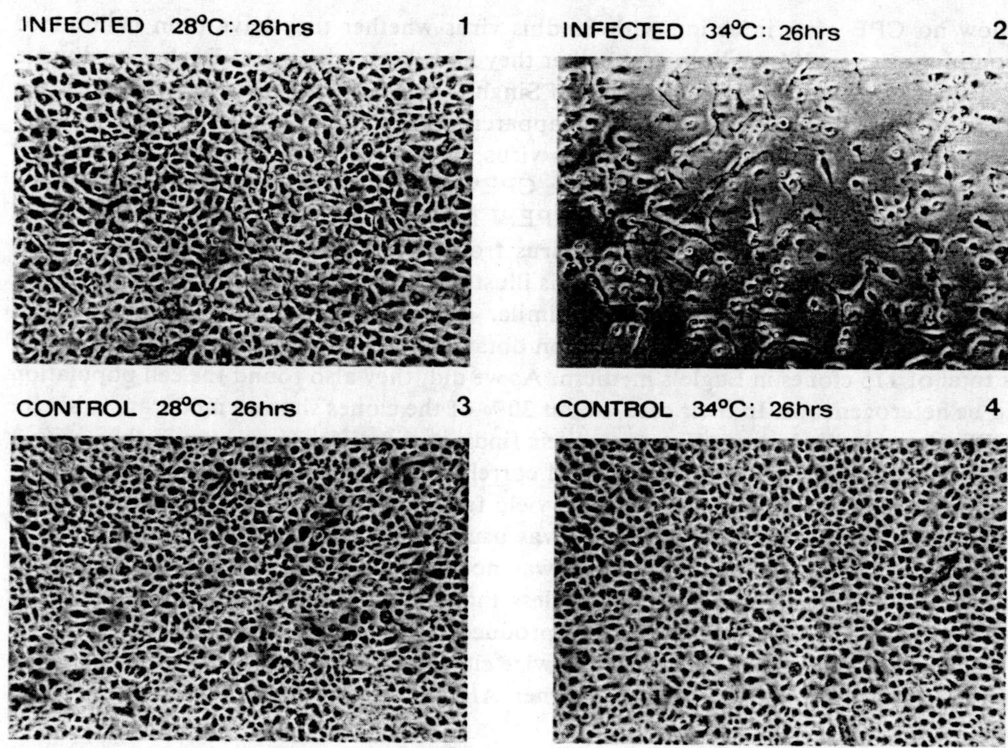

FIGURE 1. Response of a CPE⁺ *A. albopictus* cell clone to infection with Sindbis virus at 28 and 34°C. The photographs were taken 26 hr after infection. The multiplicity of infection was 20 PFU/cell. The cells are from the C-7 clone of *A. albopictus* cells[12] and were maintained in the presence of 10% fetal calf serum.

first clonally derived populations of mosquito cells were obtained.[12] These initial clones were actually obtained by Igarashi who was then working in our laboratory.

Whereas uncloned populations of Singh's *A. albopictus* cells (whether maintained in MM medium or in Eagle's medium) showed no CPE following infection with Sindbis virus (SV), clonally derived populations showed great variability.[12] In our laboratory nearly all virological studies are now done with cloned populations, either C-7 cells which have been cloned once, or C-7-10 cells, a clonal derivative of the C-7 cells. Igarashi, beginning with uncloned cells from our laboratory, obtained his C6/36 cells after two successive clonings.[13] These cells which are now used in many different laboratories are probably very similar to our C-7 cells.

Finally, the reader is referred to a recent review[14] for a listing of all mosquito cell lines which are currently available. Not all of these lines, however, have been characterized in terms of their ability to support viral replication.

III. OUTCOME OF INFECTION OF *AEDES ALBOPICTUS* CELLS WITH DIFFERENT VIRUSES

A. Family: Togaviridae
1. Genus: Alphavirus

Cells infected with either Sindbis or Semliki Forest viruses (the prototype alphaviruses) routinely produce high levels of virus ranging from 10^8 to as high as 6×10^9 PFU/mℓ.[2,6,15,16] Generally, maximal yields are obtained by 24 hr and probably even earlier in some cases.

As already noted, uncloned populations of Singh's *A. albopictus* cells generally show no CPE after infection with Sindbis virus whether they have been maintained continuously in MM medium or whether they have been adapted to Eagle's medium.

However, following the adaptation of Singh's *A. albopictus* cells to Eagle's medium and the subsequent cloning, it became apparent that the cell population was heterogeneous.[12,17] After infection with Sindbis virus, 14 out of 39 (36%) randomly selected clones showed a marked CPE, i.e., were CPE+ and 18 or 46% showed none, i.e., were CPE- and 7 or 18% showed a mild CPE.[17] No significant or consistent differences were observed in the 24 hr yields of virus from selected CPE+ and CPE- cells.[12] An example of Sindbis virus-induced CPE is illustrated in Figure 1.

Tooker and Kennedy[16] performed similar experiments with Semliki Forest virus. They began with an uncloned population obtained from our laboratory and generated a total of 115 clones in Eagle's medium. As we did, they also found the cell population to be heterogeneous. In their case, about 30% of the clones showed moderate to severe CPE after viral infection. However, their findings did differ from ours in that, with a few exceptions, they found a very good correlation between the CPE+ phenotype and the virus yield. Thus, whereas the virus yield from CPE+ clones ranged from 10^9 to 8×10^9 PFU/mℓ, that from CPE- clones was usually less than 10^8 PFU/mℓ. It is worth noting, however, that this correlation was not perfect. There were examples, for instance, of CPE+ clones which yielded less than 10^8 PFU/mℓ virus as well as clones which showed only very mild CPE but produced virus yields of 10^9 PFU/mℓ.

Note was made above of Igarashi's twice cloned C6/36 cells. These cells, following infection with chikungunya virus (another *Alphavirus*) produced high yields of virus and were CPE+.[13]

2. Genus: Pestivirus

One other virus which appears to be a togavirus has been studied in *A. albopictus* cells. This virus, termed cell fusing agent or CFA, was originally isolated from a line of *A. aegypti* cells.[21] Based on its physical and biochemical properties (spherical particle, diameter about 50 nm, ether-sensitive, infectious RNA) it was thought that the CFA may be a *Flavivirus*.[21,22] However, no serological cross-reaction could be demonstrated between the CFA and several flaviviruses which were tested. Furthermore the pattern of its structural proteins, as analyzed by polyacrylamide gel electrophoresis, was different than that typically seen with either alphaviruses or flaviviruses. There was some resemblance, however, to what has been described for the lactate dehydrogenase virus.[22] For all of these reasons, it was suggested that the CFA be assigned to the family Togaviridae, genus *Pestivirus*.[35]

When Singh's *A. albopictus* cells (uncloned) were infected with the CFA, the striking feature was cell fusion and formation of very large syncytia (Figure 2). Syncytium formation was first observed at about 60 hr after infection and reached a maximal level by about 76 hr.[21] It is worth noting that infection of these cells with type 2 dengue virus (which has been reported by other workers to cause cell fusion) has not been associated with cell fusion in our laboratory, even though reasonably high yields (i.e., 10^8 PFU/mℓ) have been obtained.[36]

B. Family: Flaviviridae

In contrast to the situation when mosquito cells are infected with alphaviruses, the cytopathic effect in *Flavivirus*-infected cells (when observed) is most often characterized by cell fusion and the appearance of large syncytia. Furthermore, this may occur with uncloned cell populations as well as with cloned cells. For example, syncytium formation was observed by Suitor and Paul[18] when they infected mosquito cells with dengue virus and by Paul et al.[19] in infections with dengue, West Nile, or Japanese encephalitis viruses. In neither case had the cells (the *A. albopictus* line of Singh) been cloned. In the report of Suitor and Paul,[18] the level of virus produced did not exceed 10^5 PFU/ml and first evidence of cell fusion was not observed until 6 days after infection.

Igarashi et al.[20] tested a number of different strains of Japanese encephalitis virus in *A. albopictus* cells and with one strain, JAOH-0566, obtained titers as high as 10^9 PFU/ml. In spite of this high yield, the cells (again an uncloned population) showed no apparent cytopathic effect. Working with his twice-cloned C6/36 *A. albopictus* cells, Igarashi observed that infection with any one of the four dengue virus serotypes, dengue types 1, 2, 3, or 4 led to clear cytopathic effect.[13] In dengue type 2 and type 4 infections, syncytia were very prominent. Infections with the other virus types caused mainly cell rounding and detachment of cells from the monolayer.

C. Family: Bunyaviridae

Relatively little work has been done with viruses of this family. Germiston virus when added to *A. albopictus* cells at high multiplicity, produced fusion from without, i.e., prior to virus replication.[23] This same virus preparation caused similar effects on several vertebrate cell lines which were tested. On the other hand, infection of the same cell line in another laboratory with Bunyamwera virus[24] was not associated with any type of cytopathic effect.

D. Family: Rhabdoviridae

Working with the C-7 clone of *A. albopictus* cells, Gillies and Stollar[25] described a highly productive infection with vesicular stomatitis virus (VSV). Yields of 10^9 PFU/ml were generally obtained and were accompanied by a very severe CPE, the beginning of which could be seen by 12 hr after infection.

IV. BIOCHEMICAL CORRELATES OF CELL KILLING IN VSV-INFECTED *AEDES ALBOPICTUS* CELLS

The preceding section documented that the response of cultured mosquito cells (even when derived from a single cell line) to infection with arboviruses can be quite varied. Obviously the outcome of infection depends on the virus with which the cells are infected. It is also quite evident that certain properties, still unknown, of the host cell play an extremely important role in determining virus yield and the fate of the infected cells.

However, very little work has been done which could help explain in biochemical terms what determines the outcome of virus infection in any given system. Why, for instance, does the *Flavivirus* infection in mosquito cells quite often lead to the formation of syncytia, whereas in infections of mammalian cells, this has not been reported? What is the critical property of the *A. albopictus* cells which determines whether or not infection with Sindbis or Semliki Forest virus will result in cell death?

The only mosquito cell system in which an attempt has been made to examine the biochemical basis of virus-induced cytopathology has been in cells infected with VSV.

Working first with intact cells (the C-7 clone of *A. albopictus* cells) it was shown that in order to produce rapid and extensive CPE in VSV-infected cells, it was necessary to incubate infected cultures in a medium containing serum rather than in a serum-free medium, and at 34°C rather than at 28°C.[26] Although cultures maintained at 28°C without serum showed little or no CPE, final virus yields were not less than in cultures showing CPE. Thus, it was possible in this system to establish conditions under which CPE was dissociated from virus production.

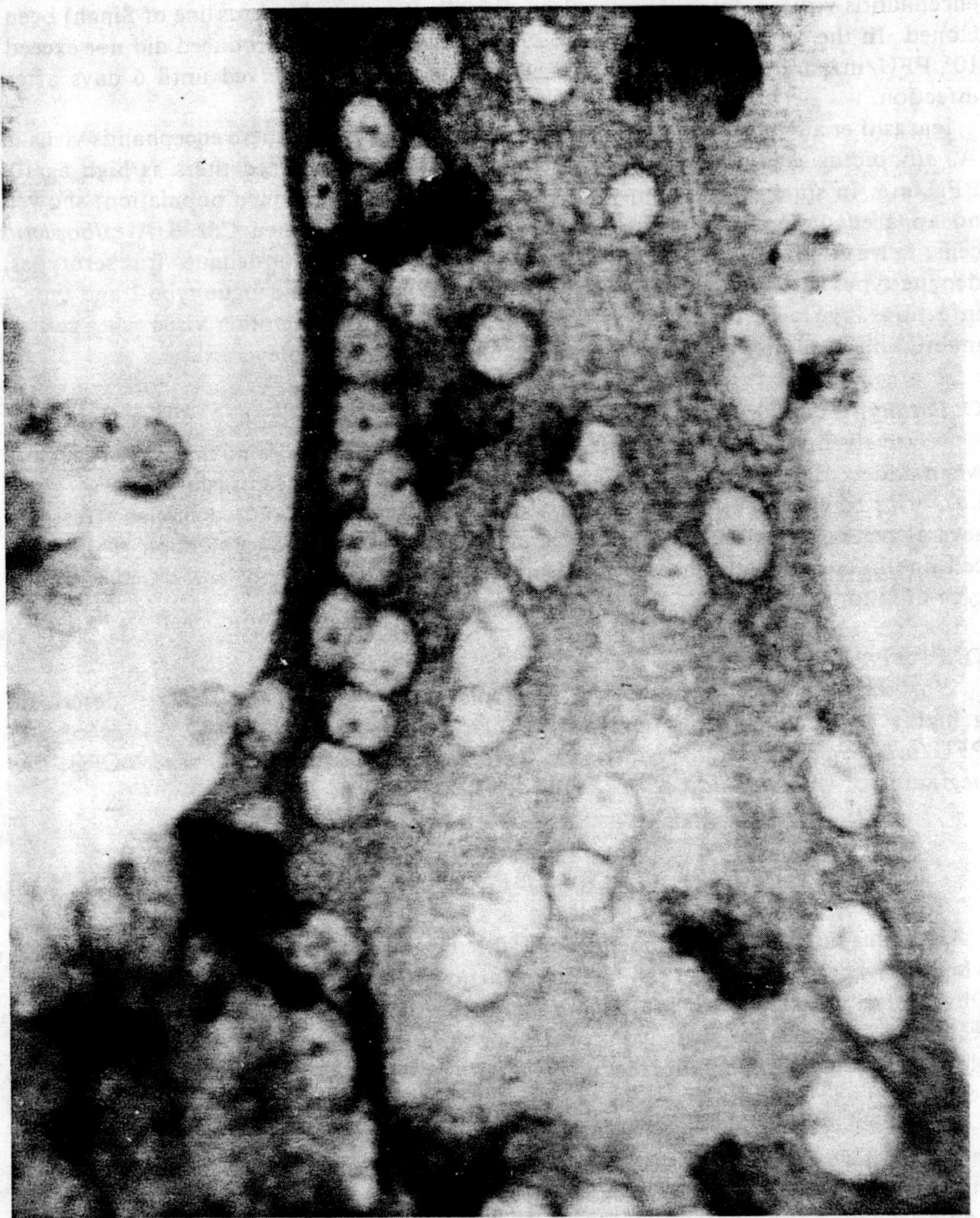

FIGURE 2.   Syncytium formation in a culture of *A. albopictus* cells infected with the cell-fusing agent (CFA). The photograph was taken 76 hr after infection. Most of the cells shown here have fused into one large syncytium. Some unfused cells are seen at the middle of the left side of the figure. (From Stollar, V. and Thomas, V. L., *Virology*, 64, 367, 1975. With permission.)

Incubation at 34°C with 10% serum also led to a marked inhibition of protein (Figure 3) and RNA synthesis in VSV-infected cells and to a dissociation of polysomes (Figure 4) to 80s monosomes.[26] This last finding is consistent with the idea that inhibition of protein synthesis under these conditions is associated with a block at the level of initiation.

In general, the rate of viral macromolecular synthesis, as measured by incorporation

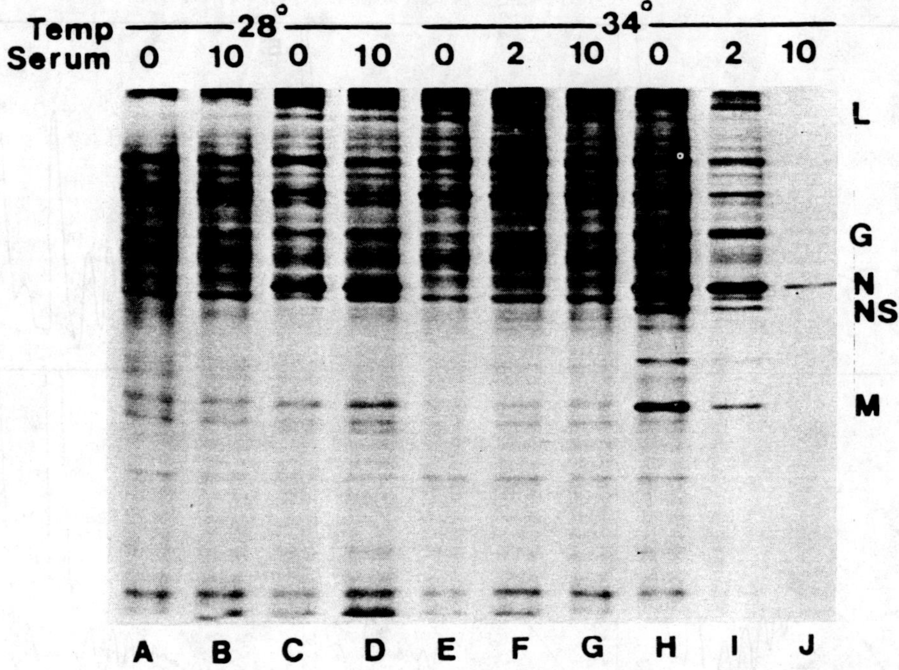

FIGURE 3.   Effect of temperature and serum concentration on protein synthesis in uninfected and VSV-infected *A. albopictus* cells. Cells were mock-infected (lanes A, B, E, F, and G) or infected (lanes C, D, H, I, and J) with VSV (10 PFU/cell) and maintained at 28 or 34°C in medium containing 0, 2, or 10% fetal calf serum. At 6 hr after infection, ³H-leucine was added and incubation was continued for 2 hr. Cell extracts were prepared and the labeled proteins were analyzed on a 10% polyacrylamide gel. L, G, N, NS, and M are the 5 VSV proteins. (From Gillies, S. and Stollar, V., *Mol. Cell Biol.*, 2, 66, 1982. With permission.)

of radioactive precursors into TCA-precipitable material, was greater in cultures incubated at 34°C rather than at 28°C. When infected cultures were maintained at 28°C in the absence of serum, treatment with low levels of actinomycin (but at sufficient concentration to cause a rapid decrease in the amount of translatable host mRNA) led to a significant enhancement of viral protein synthesis.[26] This finding suggests that under these conditions, translation of the relatively small amount of VSV mRNA present in infected cells at 28°C was limited by competition with host mRNAs for ribosomes.

To obtain more information about how VSV infection might lead to inhibition of protein synthesis, the phosphorylation patterns of proteins from infected and uninfected cells were compared.[26] The major finding of interest was that the labeling by ³²P inorganic phosphate of a protein band labeled p2, p3, which was quite prominent in uninfected cells or in infected cells maintained at 28°C (in this experiment all cultures were maintained in 10% serum), was markedly diminished when infected cells were kept at 34°C (Figure 5). The labeling of most other phosphoproteins was affected very little under these same conditions. To determine the cellular localization of the p2, p3 proteins, uninfected cells were labeled with ³²P inorganic phosphate and fractionated into polysomal and nonpolysomal fractions. The p2, p3 proteins were found to be polysome-associated[26] (Figure 5).

The findings presented above suggested that inhibition of protein synthesis in VSV-infected cells might be attributed to two things: (1) to competition by excess viral mRNA for a limiting number of ribosomes; this could account for the inhibition of host protein synthesis especially when the level of viral RNA synthesis is high, as is the case at 34°C, and (2) to cytotoxic effects resulting from the build-up of viral products, effects that for unknown reasons require the presence of serum in the medium. Such

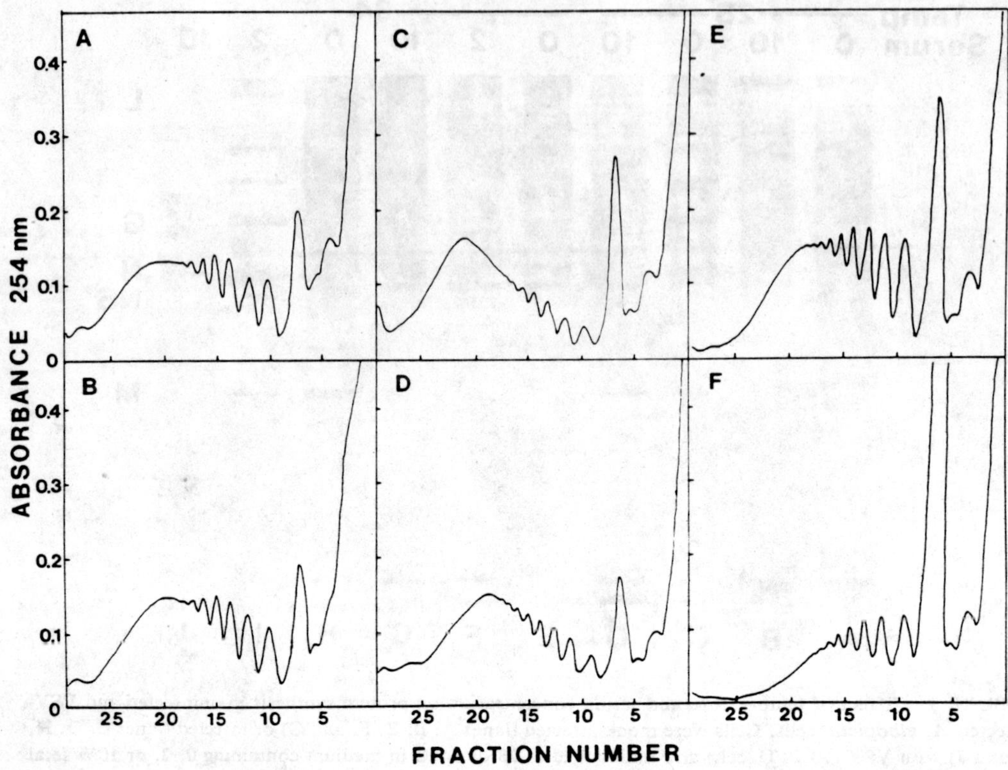

FIGURE 4.	Polysomes in uninfected and VSV-infected *A. albopictus* cells. Cytoplasmic extracts from mock- or VSV-infected *A. albopictus* cells were prepared at 6 hr after infection and analyzed by sucrose gradient centrifugation. (A) Mock-infected cells were maintained at 28°C in 10% fetal calf serum; (B) VSV-infected cells were maintained at 28°C in 10% fetal calf serum; (C) mock-infected cells were maintained at 34°C with no serum; (D) VSV-infected cells were maintained at 34°C with no serum; (E) mock-infected cells were maintained at 34°C with 10% fetal calf serum; and (F) VSV-infected cells were maintained at 34°C with 10% fetal calf serum. (From Gillies, S. and Stollar, V., *Mol. Cell Biol.*, 2, 66, 1982. With permission.)

cytotoxic effects could be associated with an inhibition of both host and viral protein synthesis. As noted above, the fact that incubation at 34°C as opposed to 28°C is associated in this system with CPE and inhibition of protein synthesis might most easily be attributed to the increased synthesis of viral macromolecules at the higher temperature. Exactly why serum is required for inhibition of protein synthesis as well as for the other effects described is still not known.

To probe further into the mechanisms involved in inhibition of protein synthesis, use was made of a cell-free translation system (S-10 fractions, 10,000 g supernatants were used for most of these experiments) prepared from *A. albopictus* cells.[27,28] The response of the extracts to the Sindbis virus 26S mRNA was used as a measure of the ability to initiate protein synthesis.

The protein synthesizing ability of cell extracts reflected what was found with whole cells[28] (Figure 6). Thus, endogenous protein synthesis in an extract prepared 6 hr after infection from VSV-infected cells maintained at 34°C in the absence of serum was only slightly less than in the extract from uninfected cells maintained at 28°C. Both extracts showed a modest increase in protein synthesis when Sindbis virus 26S mRNA was added. However, with the extract from VSV-infected cells maintained at 34°C in the presence of serum (10%), protein synthesis was drastically reduced whether measured with or without added 26S mRNA. Activity of this extract was, however, partially restored by addition of a 100,000 g supernatant fraction (S100) from uninfected cells.[28]

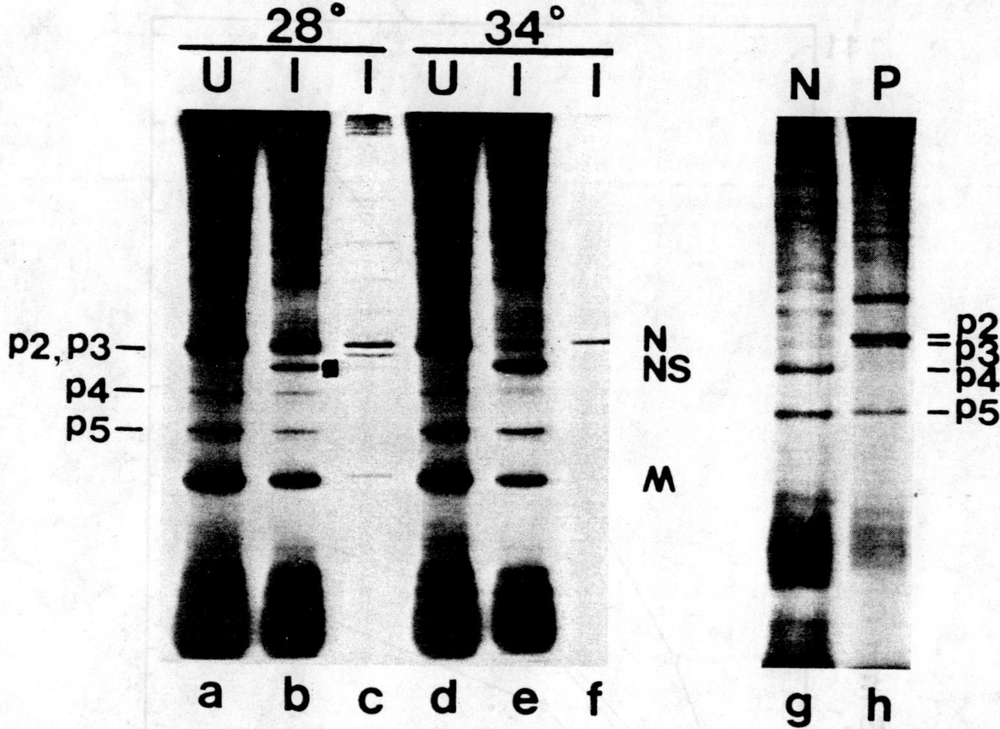

FIGURE 5.   Protein phosphorylation in uninfected and VSV-infected *A. albopictus* cells. *A. albopictus* cells were mock-infected or infected with VSV and maintained at 28 or 34°C in medium containing 10% serum. Cultures were then labeled with $^{32}PO_4$ from 4 to 8 hr after infection and the cytoplasmic phosphoproteins were analyzed by gel electrophoresis (lanes a, b, d, and e). Extracts of [^{35}S]methionine-labeled, VSV-infected cells were analyzed in parallel lanes c and f to identify the position of the viral phosphoprotein, NS. Phosphoproteins from uninfected cells were also fractionated into nonpolysomal (N) and polysomal (P) phosphoproteins (lanes g and h). These fractions were analyzed on a separate gel that was run for a shorter time. Under these conditions the phosphoproteins were resolved into two bands, p2 and p3. Phosphoprotein p1 (not shown) is detected only in cell-free extracts labeled in vitro. U is uninfected cells and I is VSV-infected cells. The symbol ■ (between lanes b and c) designates the position of the viral NS phosphoprotein. (From Gillies, S. and Stollar, V., *Mol. Cell Biol.*, 2, 66, 1982. With permission.)

This was shown by measurement of incorporation of ^{35}S methionine into TCA-precipitable material (Figure 6) as well as by monitoring the products made in vitro by polyacrylamide gel electrophoresis. The active component in the S100 fraction was nondialyzable and heat labile, and is therefore likely to be a protein. Following the fractionation of the S100 with ammonium sulfate, activity was found in the 40 to 70% saturation fraction.

Phosphorylation of the p2 and p3 proteins just described (these proteins have molecular weights of 48,000 and 46,000, respectively) was also demonstrated in vitro when active extracts were used.[28] However, an extract from VSV-infected cells maintained at 34°C with 10% serum showed decreased phosphate labeling of p2. Again, addition of an S100 fraction enhanced phosphorylation of this protein.

The effect of various inhibitors of protein synthesis on the phosphorylation of p2 was tested.[28] Inhibitors of elongation had no effect, but the addition of pactomycin which blocks initiation by preventing the joining of ribosomal subunits did inhibit phosphorylation of p2. This suggests that the biochemical lesion responsible for shutting off protein synthesis in VSV-infected cells may also affect the joining of ribosomal subunits. Analysis of initiation complexes from infected cells maintained at 34°C with or without serum did indeed show that in the case of cells incubated with serum, for-

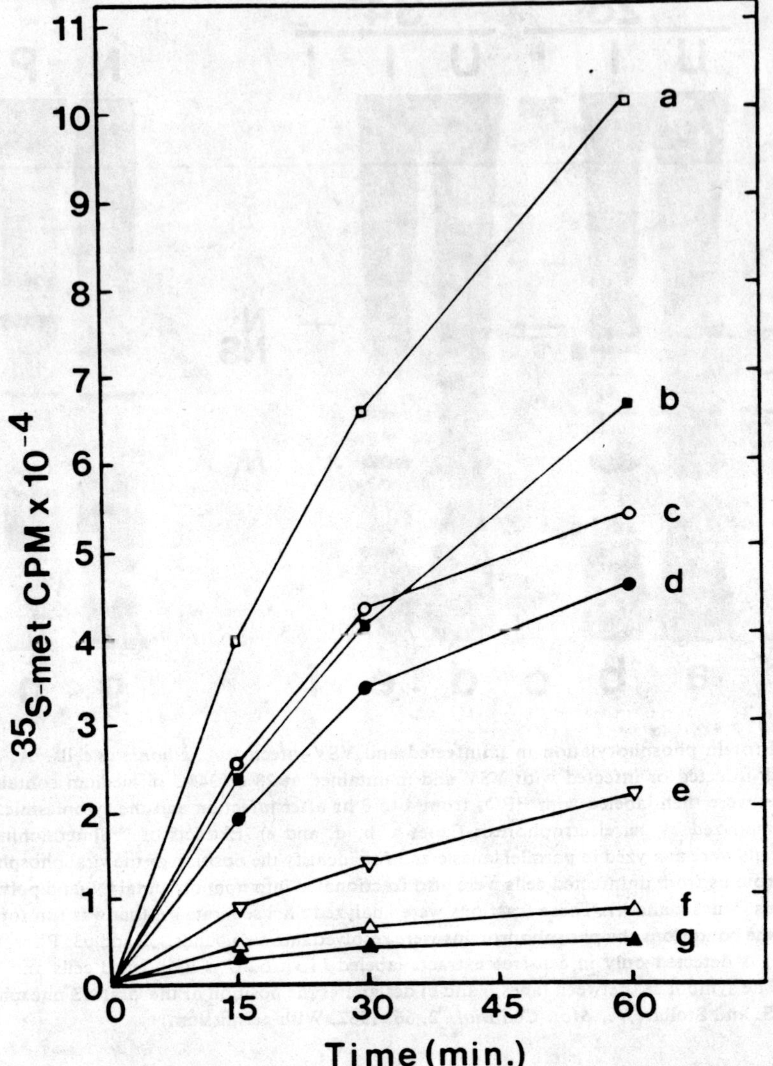

FIGURE 6.   In vitro protein synthesis in extracts prepared from uninfected and VSV-infected *A. albopictus* cells. S-10 extracts were incubated at 28°C under conditions for protein synthesis in the absence or presence of SV 26S RNA. The designations S-10/0 and S-10/10 refer to S-10 extracts prepared from cells maintained in the absence and presence of fetal calf serum (10%), respectively. The total reaction volume was 20 μl. [³⁵S]methionine incorporation was measured in 2 μl aliquots at the indicated times. Extracts were prepared from uninfected cells maintained at 28°C (a, b) and from VSV-infected cells maintained at 34°C (c, d, e, f, g). (a) Uninfected cell S-10/10 extract plus added SV 26S RNA; (b) uninfected cell S-10/10 extract, no added RNA; (c) S-10/0 extract from VSV-infected cells plus SV 26S RNA; (d) S-10/0 extract from VSV-infected cells, no added RNA; (e) S-10/10 extract from VSV-infected cells plus SV 26S RNA and S-100 from uninfected cells (4 μl of S-100/20 μl total reaction volume); (f) S-10/10 extract from VSV-infected cells plus SV 26S RNA; and (g) S-10/10 extract from VSV-infected cells, no added RNA. (From Gillies, S. and Stollar, V., *Virology*, 2, 1174, 1982. With permission.)

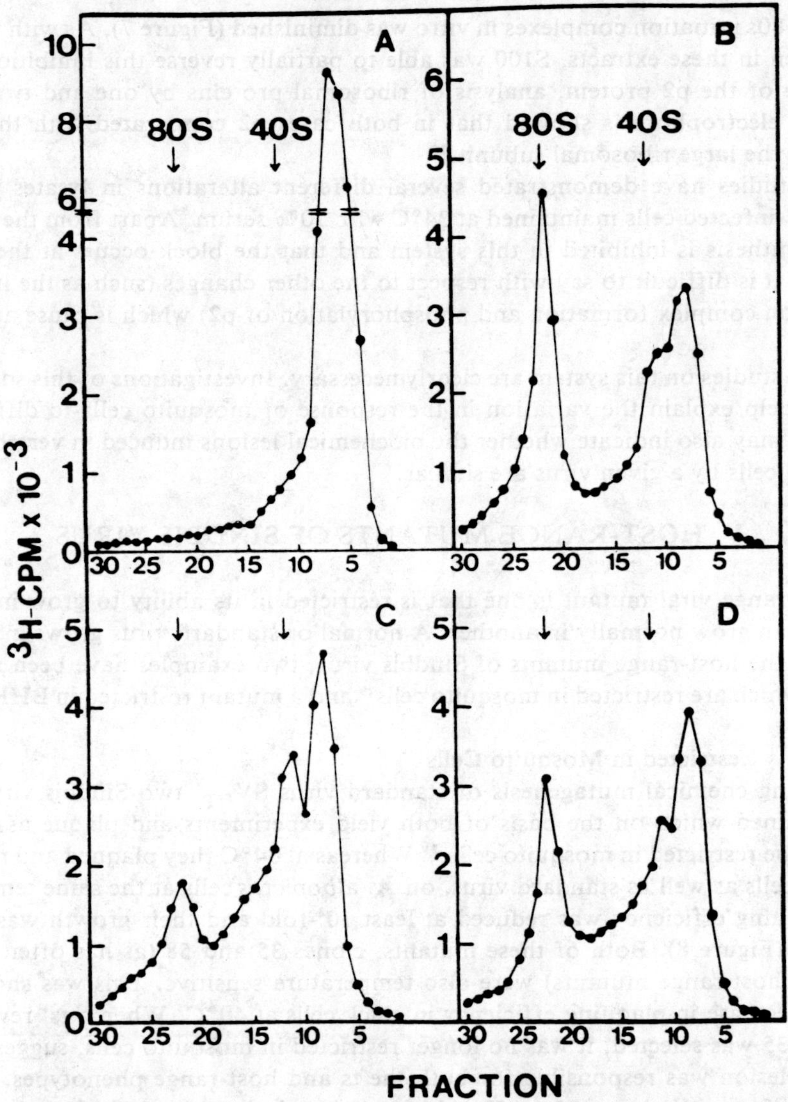

FIGURE 7.   Labeling of 40S and 80S initiation complexes with VSV 12 to 18S
mRNA in extracts from VSV-infected cells. S-10 extracts were prepared from VSV-
infected cells maintained at 34°C. The binding of [³H]uridine-labeled VSV 12 to
18S mRNA to 40S and 80S initiation complexes was carried out in the presence of
anisomycin (400 μg/mℓ), a compound which prevents elongation but permits the
formation of initiation complexes. After centrifugation of the reaction mixtures
through a 7 to 30% sucrose gradient, the [³H]uridine-labeled RNA was precipitated
from the gradient fractions with cetyltrialkyl ammonium bromide. (A) Binding of
RNA in an S-10/0 extract in the presence of 1 m*M* aurintricarboxylic acid (ATA),
ATA is known to prevent the binding of mRNA to the 40S complex and thus also
to the 80S initiation complex; (B) binding of RNA in an S-10/0 extract in the absence of ATA; (C) binding of RNA in an S-10/10 extract; and (D) binding of RNA
in an S-10/10 extract supplemented with S-100 from uninfected cells (10 μℓ of S-
100/50 μℓ reaction). (From Gillies, S. and Stollar, V., *Virology*, 2, 1174, 1982. With
permission.)

mation of 80s initiation complexes in vitro was diminished (Figure 7). As with the other effects seen in these extracts, S100 was able to partially reverse this inhibition. As to the nature of the p2 protein, analysis of ribosomal proteins by one and two dimensional gel electrophoresis showed that in both cases p2 comigrated with the largest protein of the large ribosomal subunit.[28]

These studies have demonstrated several different alterations in lysates prepared from VSV-infected cells maintained at 34°C with 10% serum. Apart from the fact that protein synthesis is inhibited in this system and that the block occurs at the level of initiation, it is difficult to say with respect to the other changes (such as the inhibition of initiation complex formation and phosphorylation of p2) which is cause and which is effect.

Further studies on this system are clearly necessary. Investigations of this sort should not only help explain the variation in the response of mosquito cells to different viruses, but may also indicate whether the biochemical lesions induced in vertebrate and arthropod cells by a given virus are similar.

V. HOST-RANGE MUTANTS OF SINDBIS VIRUS

A host-range viral mutant is one that is restricted in its ability to grow in one cell type but can grow normally in another. A normal or standard virus grows in both cell types. Of the host-range mutants of Sindbis virus, two examples have been reported: mutants which are restricted in mosquito cells[29] and a mutant restricted in BHK cells.[30]

A. Mutants Restricted in Mosquito Cells

Following chemical mutagenesis of standard virus SV_{STD}, two Sindbis virus clones were obtained which on the basis of both yield experiments and plaque assays were shown to be restricted in mosquito cells.[29] Whereas at 34°C they plaqued and replicated on chick cells as well as standard virus, on *A. albopictus* cells at the same temperature their plaquing efficiency was reduced at least 10^4-fold and their growth was severely restricted (Figure 8). Both of these mutants, clones 35 and 58 (as has often been the case with host-range mutants) were also temperature sensitive. This was shown by a marked decrease in plaquing efficiency in chick cells at 40°C. When a ts⁺ revertant of SV clone 35 was selected, it was no longer restricted in mosquito cells, suggesting that the same lesion was responsible for both the ts and host-range phenotypes. The fact that RNA from SV clones 35 or 58, which produced plaques on chick embryo fibroblasts (CEF), was unable to form plaques on *A. albopictus* cell monolayers (RNA from SV_{STD} forms plaques on both cell types) indicated that the block to replication in *A. albopictus* cells of the mutant viruses was not at the level of adsorption.

There were two lines of evidence indicating that the defect in SV clones 35 and 58 was at an early stage in replication. First, relative to viral RNA synthesis induced by SV_{STD}, ³H-uridine incorporation into viral RNA (actinomycin-resistant RNA synthesis) was 90% or more reduced under nonpermissive conditions, i.e., in CEF at 40°C or in *A. albopictus* cells at 34°C. Second, the results of complementation tests at 40°C in CEF with SV clones 35 and 58 and known ts mutants of SV showed that both clone 35 and 58 belonged to complementation group F, one of the four known complementation groups to which RNA⁻ mutants have been assigned.

Although there is at this time no firm evidence pointing to a host component in the Sindbis virus RNA polymerase, it is possible that as with poliovirus,[31] an interaction between a specific host component and a virus-coded protein is necessary for the assembly of a functional polymerase. If so, one might speculate that the defect in the Sindbis virus host-range mutants, clones 35 and 58, involves a viral protein which

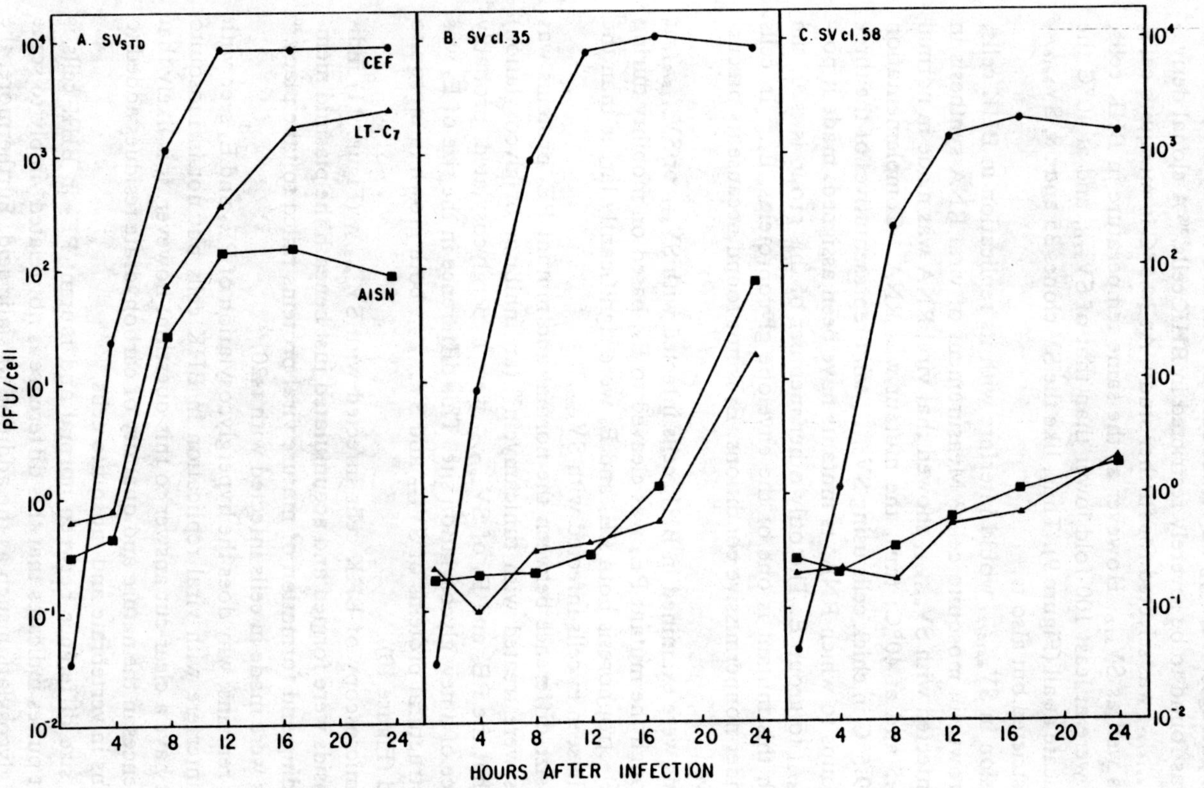

FIGURE 8.   Replication of Sindbis virus host-range mutants in cultures of chick embryo and mosquito cells at 34.5°C. Cultures of CEF (•), *A. albopictus* clones LT-C7 (▲), and AIS-N (■) were infected (moi = 10) with SV_{STD} (A), SV clone 35 (B), or SV clone 58 (C) and incubated at 34.5°C. Samples of medium were taken at the indicated times and assayed for infectious virus by plaque assay on CEF cells at 34.5°C. The results are expressed as the yield of virus released per cell. AIS-N cells are CPE⁻ while LT-C7 cells are CPE⁺. (From Kowal, K. J. and Stollar, V., *Virology*, 114, 140, 1981. With permission.)

interacts with this hypothetical host component. The defect in this viral protein would be such that at 34°C it is able to interact properly with the host component from CEF, but not from *A. albopictus* cells. Presumably at 40°C, because of its thermal lability, the interaction with the host component from CEF would be impaired.

B. A Mutant of Sindbis Virus Able to Replicate in Mosquito but not Vertebrate Cells

Following 15 serial passages of SV_{STD} on *A. albopictus* cells at 34.5°C, the resultant virus stock SV_{ap15}, was found to be severely restricted in BHK cells.[30] A clonal derivative of this stock, $SV_{ap15/21}$, was selected for further study. At 34.5°C in mosquito cells, $SV_{ap15/21}$ grew just as well as SV_{STD}. However at the same temperature in BHK cells, yields of the mutant were at least 100-fold lower than that of SV_{STD}, and at 40°C the mutant failed to replicate at all (Figure 9). Thus, like the SV clones 35 and 58, $SV_{ap15/21}$ was not only host-restricted but also ts.

What kind of a lesion in $SV_{ap15/21}$ would interfere with its replication in BHK cells, but still permit its growth in mosquito cells? Measurement of viral RNA synthesis in BHK or chick cells infected with $SV_{ap15/21}$ showed that viral RNA was made in normal amounts both at 34.5 and at 40°C; thus, the mutant was RNA^+. Complementation tests carried out at 39.5°C in chick cells using $SV_{ap15/21}$ and representatives of the three complementation groups to which RNA^+ ts mutants have been assigned, made it possible to assign $SV_{ap15/21}$ to group E. The only other member of this group is ts20; the affected protein with this mutant is one of the envelope glycoproteins, E_2.[32] In cells infected with ts20 under nonpermissive conditions, E_2 is not found because its precursor, PE_2, is not cleaved.

When viral proteins were examined in BHK cells infected with SV_{STD} or $SV_{ap15/21}$, it was found that although the mutant PE_2 was cleaved to E_2, based on mobility during polyacrylamide gel electrophoresis both PE_2 and E_2 were significantly larger than the corresponding protein seen in cells infected with SV_{STD}.[30]

The fact that the size difference between the normal and mutant PE_2 proteins was abolished when cells were treated with tunicamycin[30] (an inhibitor of glycosylation) strongly suggested that the PE_2 and E_2 of $SV_{ap15/21}$ were hyperglycosylated, probably due to the appearance of a new glycosylation site. This difference in the size of E_2 was also seen when the structural proteins of SV_{STD} and $SV_{ap15/21}$, both grown in mosquito cells, were compared (Figure 10).

Finally, electron microscopy of BHK cells infected with $SV_{ap15/21}$ (Figure 11) indicated that nucleocapsids were formed and accumulated just beneath the plasmid membrane, but that budding and formation of mature viral proteins failed to take place.[30] Similar observations were made in cells infected with ts20.[33]

The question then remains, why does the hyperglycosylation of PE_2 and E_2 seen with $SV_{ap15/21}$ infections interfere with viral replication in BHK cells but not in mosquito cells? We do not yet have a clear-cut answer to this question, however, it is likely that it is related to differences in the nature and quantity of carbohydrate residues added to the viral glycoproteins in vertebrate and mosquito cells.[5,7,34]

The fact that the size difference between normal and mutant PE_2 in BHK cells is evident even in short pulses indicates that this difference is not related simply to some late modification in glycosylation such as the addition of sialic acid. Furthermore, the difference in the size of E_2 was seen with virus grown in *A. albopictus* cells; the glycoproteins from a virus grown in mosquito cells do not contain sialic acid or indeed any complex oligosaccharides.[5,7,34]

The best explanation then for the ability of $SV_{ap15/21}$ to replicate in mosquito cells, even though it is restricted in BHK cells, lies with the fact that the carbohydrate moieties of viral glycoproteins made in mosquito cells are different from those made in vertebrate cells.[5,7,34] It is possible, for instance, that the addition of a new glycosylation

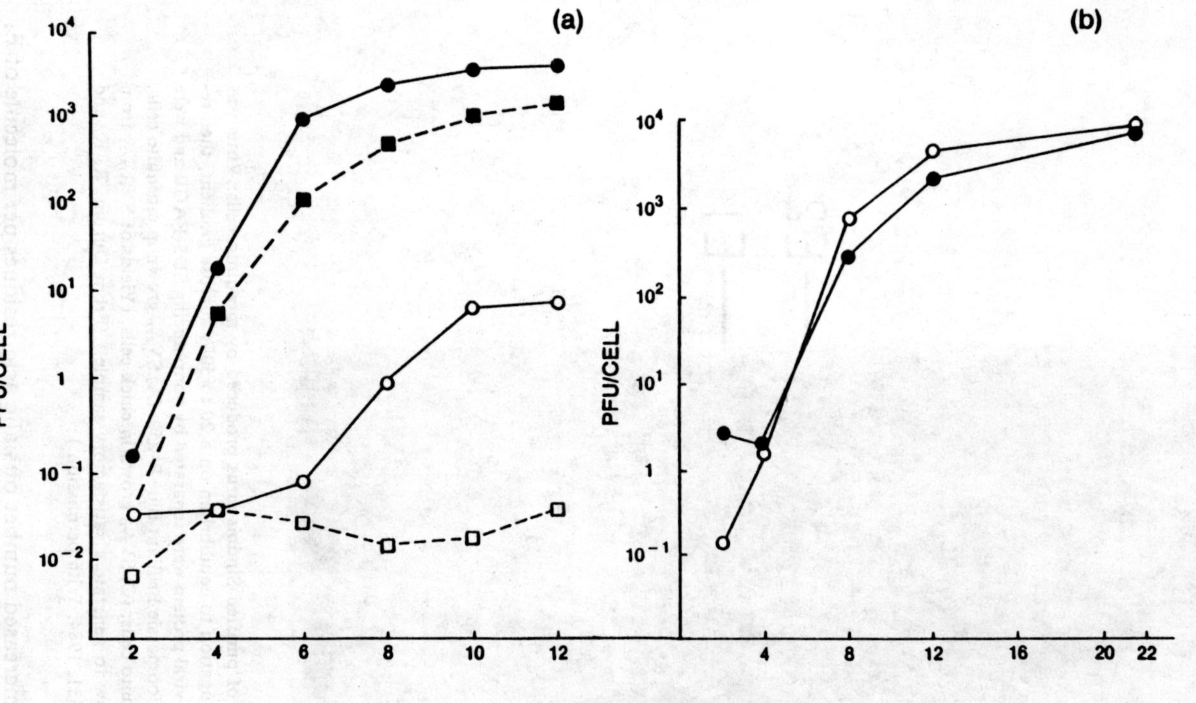

FIGURE 9.   Replication of $SV_{ap15/21}$ and SV_{STD} in BHK and mosquito cells. Cells were infected at an input multiplicity of 10 PFU/ cell. BHK cells were incubated at either 34.5 or 40°C, *A. albopictus* cells at 34.5°C. At the indicated times postinfection, samples were taken and titers were determined by plaque assay on chick cells at 34.5°C. (a) BHK cells; (b) *A. albopictus* cells; (O) $SV_{ap15/21}$ at 34.5°C; (□) $SV_{ap15/21}$ at 40°C; (•) SV_{STD} at 34.5°C; and (■) SV_{STD} at 40°C. (From Durbin, R. K. and Stollar, V., *Virology*, 135, 331, 1984. With permission.)

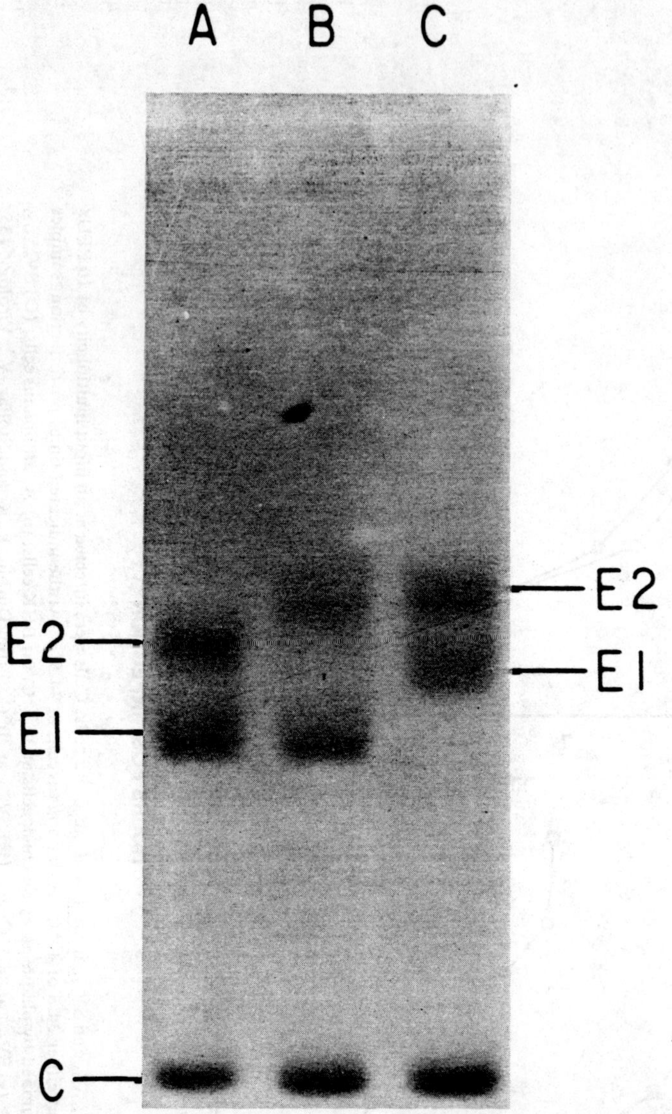

FIGURE 10.   SDS-PAGE of purified Sindbis virus produced by mosquito cells. Virus was pelleted, purified by centrifugation to equilibrium on a 20 to 50% sucrose gradient, then re-pelleted. Six micrograms of viral protein were separated by nonreducing SDS-PAGE and were visualized by staining with Coomassie Brilliant Blue R-250. (A) SV_{STD} grown in mosquito cells; (B) $SV_{ap15/21}$ grown in mosquito cells; (C) SV_{STD} grown in chick cells. (Yields of $SV_{ap15/21}$ from vertebrate cells were too low to generate a satisfactory sample.).(From Durbin, R. K. and Stollar, V., *Virology*, 135, 331, 1984. With permission.)

site on PE_2 and E_2 and the increased number of sialic acid residues per molecule of E_2 could result in electrostatic repulsion and decreased affinity between viral peplomers in infected vertebrate cells, thus interfering with virus maturation. Alternatively, the additional carbohydrate on the E_2 of $SV_{ap15/21}$ might interfere sterically with the interaction between viral structural units.

Although it seems clear that the restriction of $SV_{ap15/21}$ in vertebrate cells is associated with hyperglycosylation of PE_2 or E_2, it is not at all evident yet whether the host-determined differences in the oligosaccharides of E_2 that are responsible for the host-

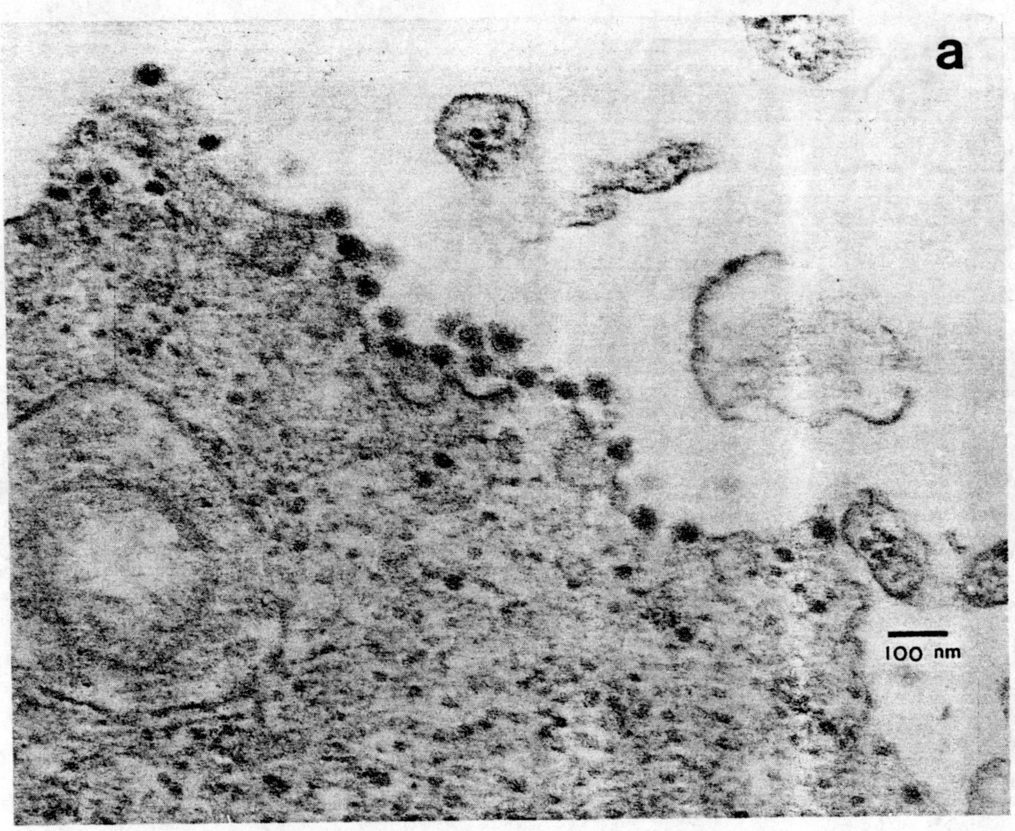

FIGURE 11. Electron micrographs of infected chick cells. Secondary cultures of chick cells were infected with 100 PFU of SV_{STD} (a) or $SV_{ap15/21}$ (b) per cell and incubated at 34.5°C. At 7.5 hr p.i., cells were fixed and prepared for electron microscopy. Bar = 100 nm. (From Durbin, R. K. and Stollar, V., *Virology*, 135, 331, 1984. With permission.)

range effect are only quantitative in nature or whether they may also be related to the type of sugar residues added to E_2 in the different host cells. No matter what the explanation, the association of a host-specific defect in viral maturation with hyperglycosylation of a viral structural protein highlights the importance of host-specific glycosylation patterns in determining the host-range of a virus. Because the glycosylation patterns differ so widely between arthropod and vertebrate cells, such a relationship between host-determined glycosylation patterns and host-range may be especially important with respect to arthropod-borne viruses.

VI. CONCLUSION

The variety of possible interactions between different arthropod-borne viruses and mosquito cells provides fascinating systems for study. Knowledge gained from such systems may lead to useful insights concerning how arthropod-borne viruses replicate in and are transmitted by mosquitoes. In addition, the careful study of host-range mutants of arthropod-borne viruses such as those described here should ultimately be of great help in defining those host cell functions and proteins required for viral replication. Host-range mutants of arboviruses are of special interest in this respect because

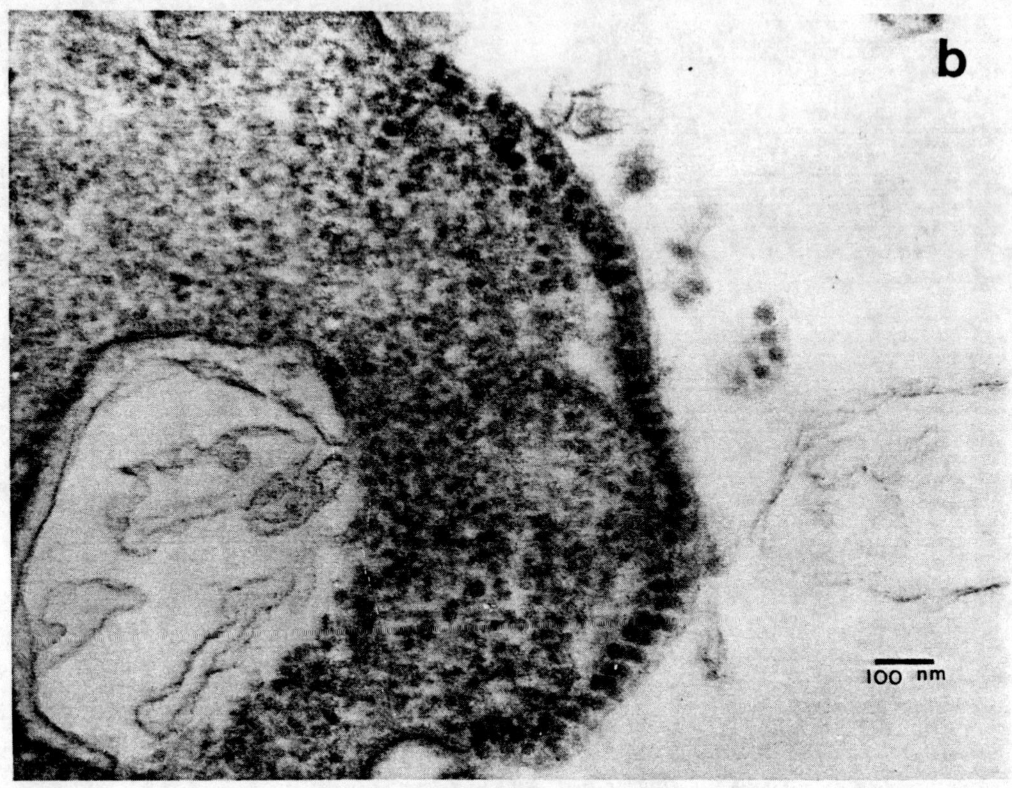

FIGURE 11b

relative to other viruses of humans and domestic animals, their natural hosts are so widely separated phylogenetically.

ACKNOWLEDGMENTS

Work in my laboratory has been supported by Public Health Service grant AI-11290 from the National Institute of Allergy and Infectious Diseases, by the U.S.-Japan Medical Science Program through Public Health Service grant AI-05920, and by an Institutional National Research Service Award (CA-09069) from the National Cancer Institute.

REFERENCES

1. Singh, K. R. P., Cell cultures derived from larvae of *Aedes albopictus* (Skuse) and *Aedes aegypti* (L), *Curr. Sci.*, 36, 506, 1967.
2. Stevens, T. M., Arbovirus replication in mosquito cell lines (Singh) grown in monolayer or suspension culture, *Proc. Soc. Exp. Biol. Med.*, 134, 356, 1970.
3. Stollar, V., Togaviruses in cultured arthropod cells, in *The Togaviruses, Biology, Structure, Replication*, Schlesinger, R. W., Ed., Academic Press, New York, 1980, 583.
4. Schlesinger, R. W., Introduction, in *The Togaviruses, Biology, Structure, Replication*, Schlesinger, R. W., Ed., Academic Press, New York, 1980, chap. 1.

5. Stollar, V., Stollar, B. D., Koo, R., Harrap K. A., and Schlesinger, R. W., Sialic acid contents of Sindbis virus from vertebrate and mosquito cells. Equivalence of biological and immunological viral properties, *Virology*, 69, 104, 1976.
6. Luukkonen, A., Kääriäinen, L. and Renkonen, O., Phospholipids of Semliki forest virus grown in cultured mosquito cells, *Biochim. Biophys. Acta*, 450, 109, 1976.
7. Luukkonen, A., Von Bonsdorff, C.-H., and Renkonen, O., Characterization of Semliki forest virus grown in mosquito cells. Comparison with the virus from hamster cells, *Virology*, 78, 331, 1977.
8. Arthropod cell cultures and their application to the study of viruses, in *Current Topics in Microbiology and Immunology*, Vol. 55, Weiss, E., Ed., Springer-Verlag, New York, 1971, chap. 5 to 7.
9. Singh, K. R. P., Growth of arboviruses in arthropod tissue culture, in *Advances in Virus Research*, Vol. 17, Smith, K. M., Lauffer, M. A., and Bang, F. B., Eds., Academic Press, New York, 1972, 187.
10. Mussgay, M., Enzmann, P. J., Weiland, E., and Horzinek, M. C., Growth cycle of arboviruses in vertebrate and arthropod cells, *Prog. Med. Virol.*, 19, 258, 1975.
11. Mitsuhashi, J. and Maramorosch, K., Leafhopper tissue culture: embryonic, nymphal and imaginal tissues from aseptic insects, *Contrib. Boyce Thompson Inst.*, 22, 435, 1964.
12. Sarver, N. and Stollar, V., Sindbis virus-induced cytopathic effect in clones of *Aedes albopictus* (Singh) cells, *Virology*, 80, 390, 1977.
13. Igarashi, A., Isolation of a Singh's *Aedes albopictus* cell clone sensitive to dengue and chikungunya viruses, *J. Gen. Virol.*, 40, 531, 1978.
14. Kurtti, T. J. and Munderloh, U. G., Mosquito cell culture, in *Advances in Cell Culture*, Vol. 3, Maramorosch, K., Ed., Academic Press, New York, 1984, 259.
15. Riedel, B. and Brown, D. T., Novel antiviral activity found in the media of Sindbis virus persistently infected mosquito *(Aedes albopictus)* cell cultures, *J. Virol.*, 29, 51, 1979.
16. Tooker, P. and Kennedy, S. I. T., Semliki forest virus multiplication in clones of *Aedes albopictus* cells, *J. Virol.*, 37, 589, 1981.
17. Sarver, N., Sindbis Virus-Induced Killing of *Aedes albopictus* Cells: Properties of Virus-Induced Cell Injury and Inhibition by Virazole, Ph.D. thesis, Rutgers University, New Brunswick, N. J., 1978.
18. Suitor, E. C., Jr. and Paul, F. J., Syncytia formation of mosquito cell cultures mediated by type 2 dengue virus, *Virology*, 38, 482, 1969.
19. Paul, S. D., Singh, K. R. P., and Bhat, U. K. M., A study on the effect of arboviruses on cultures from *Aedes albopictus* cell line, *Indian J. Med. Res.*, 57, 339, 1969.
20. Igarashi, A., Sasao, F., Wungkobkiat, S., and Fukai, K., Growth of Japanese encephalitis virus in established lines of mosquito cells, *Biken J.*, 16, 17, 1973.
21. Stollar, V. and Thomas, V. L., An agent in the *Aedes aegypti* cell line (Peleg) which causes fusion of *Aedes albopictus* cells, *Virology*, 64, 367, 1975.
22. Igarashi, A., Harrap, K. A., Casals, J., and Stollar, V., Morphological, biochemical and serological studies on a viral agent (CFA) which replicates in and causes fusion of *Aedes albopictus* (Singh) cells, *Virology*, 74, 174, 1976.
23. Djinawi, N. K. and Olson, L. C., Cell fusion induced by Germiston and Wesselbron viruses, *Arch. Gesamte Virusforsch.*, 43, 144, 1973.
24. Newton, S. E., Short, N. J., and Dalgarno, L., Bunyamwera virus replication in cultured *Aedes albopictus* (mosquito) cells: establishment of a persistent viral infection, *J. Virol.*, 38, 1015, 1981.
25. Gillies, S. and Stollar, V., The production of high yields of infectious vesicular stomatitis virus in *Aedes albopictus* cells and comparisons with replication in BHK-21 cells, *Virology*, 107, 509, 1980.
26. Gillies, S. and Stollar, V., Conditions necessary for inhibition of protein synthesis and cytopathic effect in *Aedes albopictus* cells infected with vesicular stomatitis virus, *Mol. Cell Biol.*, 2, 66, 1982.
27. Gillies, S. and Stollar, V., Translation of vesicular stomatitis and Sindbis virus mRNAs in cell-free extracts of *Aedes albopictus* cells, *J. Biol. Chem.*, 256, 13188, 1981.
28. Gillies, S. and Stollar, V., Protein synthesis in lysates of *Aedes albopictus* cells infected with vesicular stomatitis virus, *Virology*, 2, 1174, 1982.
29. Kowal, K. J. and Stollar, V., Temperature-sensitive host-dependent mutants of Sindbis virus, *Virology*, 114, 140, 1981.
30. Durbin, R. K. and Stollar, V., A mutant of Sindbis virus with a host-dependent defect in maturation associated with hyperglycosylation of E2, *Virology*, 135, 331, 1984.
31. Dasgupta, A., Zabel, P., and Baltimore, D., Dependence of the activity of the poliovirus replicase on a host cell protein, *Cell*, 19, 423, 1980.
32. Jones, K. J., Waite, M. R. F., and Bose, H. R., Cleavage of a viral envelope precursor during the morphogenesis of Sindbis virus, *J. Virol.*, 13, 809, 1974.
33. Brown, D. T. and Smith, J. F., Morphology of BHK-21 cells infected with Sindbis virus temperature-sensitive mutants in complementation groups D and E, *J. Virol.*, 10, 526, 1975.

34. Hsieh, P. and Robbins, P. W., Regulation of asparagine-linked oligosaccharide processing. Oligosaccharide processing in *Aedes albopictus* mosquito cells, *J. Biol. Chem.*, 259, 2375, 1984.

35. Porterfield, J. S., Casals, J., Chumakov, M. P., Gaidamovich, S. Ya., Hannoun, C., Holmes, I. H., Horzinek, M. C., Mussgay, M., Oker-Blom, N., Russell, P. K.; and Trent, D. W., Togaviridae, *Intervirology*, 9, 129, 1978.

36. Ryan, T. and Cleaves, G., Unpublished information.

Chapter 19

RHABDOVIRUSES IN *DROSOPHILA* CELLS

S. Dezélée, F. Wyers, D. Blondel, A. M. Petitjean, D. Contamine, F. Bras,
and D. Teninges

TABLE OF CONTENTS

I.   Introduction. Why Rhabdoviruses in *Drosophila* Cells?..........................112
     A.   The Symptom of Sensitivity to Carbon Dioxide............................112
     B.   The Viral Carrier State 112
     C.   The Sensitivity to Refractory Genes....................................113

II. Rhabdovirus (VSV, Piry, and Sigma) Multiplication in *Drosophila*
     Cells...115
     A.   Characteristics of *Drosophila* Cell Lines Used........................116
     B.   Infection Conditions 116
     C.   Comparative Growth Cycles of Rhabdoviruses.............................117

III. Persistent Infection in Insect Cells..119
     A.   Differences Between the Mechanisms of Persistent Infection
          in Insect and Vertebrate Cells..119
     B.   Compared Synthesis of Viral Components in *Drosophila*
          and Vertebrate Cells..121
          1.   VSV Protein Synthesis..121
          2.   VSV RNA Synthesis..124
               a.   VSV Transcription in *Drosophila* Cells....................124
               b.   Synthesis and Localization of VSV Leader RNA
                    in *Drosophila* Cells 125

IV. *Drosophila* Genes Which Induce Modifications in the Multiplication
     of Sigma and Piry Viruses: Refractory Genes................................126
     A.   Historical 126
     B.   Expression in Continuous Cell Lines In Vitro 128

V.   Conclusion 129

References 130

I. INTRODUCTION: WHY RHABDOVIRUSES IN *DROSOPHILA* CELLS?

Arthropod vectors of viral diseases are generally not affected in their lifespan or selective fitness by the multiplication in their tissues of the viruses they transmit. Arthropod cells cultivated in vitro exhibit the same capacity to survive infection by viruses which are lethal to vertebrate cells. Viruses belonging to groups as different in their replication strategies as Toga, Rhabdo, or Bunyaviridae show no cytopathic effect in cells from many different arthropod species, but rather they initiate a persistent infection characterized by a constant shedding of virus.

Drosophila is neither a biting insect, nor a vector for any dangerous viral disease; still, upon infection by toga or rhabdoviruses, *Drosophila* cells express the same type of persistent infection as other arthropod cells.[1-4] The mechanisms by which these cells are able to both support and control virus multiplication within limits compatible with their own survival is still a challenging enigma. Facing the problem of unraveling the biochemical functions involved in the general process of virus multiplication control among the thousands of functions of an arthropod cell, the choice of an organism extensively explored by both genetic and biochemical approaches is a help. Presently, in spite of the large amount of work performed on several vector species in these fields, the acquisition rate of fundamental knowledge and experimental means still favors *Drosophila*. The first virus described as a natural parasite of *Drosophila* (the sigma virus) happened to be a rhabdovirus.[5,6] Rhabdoviridae are ubiquitous bullet shaped viruses isolated from extremely diverse hosts: vertebrates, invertebrates, and plants. Some multiply in both vertebrates and invertebrates and some multiply in both insects and plants. The rhabdoviruses are divided into two subgroups, the vesiculoviruses with vesicular stomatitis virus (VSV) as a prototype and the lyssaviruses whose prototype is rabies.[7] The vesiculoviruses are the most frequently found in insects. The sigma virus has a very narrow host range. Aside from *Drosophila,* it was found to replicate in some mosquito species[8] but never recognized as possible hosts any of the vertebrate in vitro systems offered.[9] Several reviews have described in detail the relations of sigma virus with its host.[10-12] Some of the most striking features of this association are the following.

A. The Symptom of Sensitivity to Carbon Dioxide

CO_2 has a mild narcotic effect on insects. When returned to air, normal insects recover very fast, while flies infected by sigma virus remain paralyzed and die consequently. This symptom has been correlated with the multiplication of the virus in the thoracic ganglia[13] and was used to detect and assay the sigma virus in *Drosophila*.[14] Other rhabdoviruses from vertebrates were tested in *Drosophila* and many of them induced a similar symptom.[15-17] The CO_2 sensitivity is not limited to this insect; a variety of rhabdoviruses including the sigma virus were shown to induce it in mosquitoes.[8] This suggests that many other insect species infected with viruses able to multiply in their nerve cells might express such a symptom, a fact which could be useful in epidemiological investigations.

B. The Viral Carrier State

The sigma virus is very widespread in natural *Drosophila* populations. The carrier state can be expressed at two levels, somatic and germinal, somatic because the tissues of infected flies keep shedding virus throughout the life of the insect, and germinal because the hereditary transmission via male or female germ cells is the only known propagation process of the sigma virus in wild populations. The transmission pattern is not Mendelian, hence the viral genetic information is not linked to the host chromosomes.[18] It seems to result from an equilibrium between the rate of virus replication and germ cell division. This equilibrium can be disturbed by many factors from both host and viral origin. For example, sigma virus strains selected for high yields are generally poorly transmitted, as they induce partial sterility and the surviving progeny contains little or no virus.[11] Symmetrically temperature sensitive mutants affected in early functions are not transmitted at nonpermissive temperature and some alleles of host genes which restrict sigma virus replication also interrupt its hereditary transmission.[19] Studies of defective sigma virus strains and of temperature sensitive mutants for late functions allowed the deduction that the production of infectious virions is not indispensable for the maintenance of this carrier state.[20] *Drosophila* flies inoculated with other vesiculoviruses expressed a persistent infection in their somatic cells, but no hereditary transmission was observed. However, in other insects such as sandflies, VSV could invade the germ line cells and was transmitted to the progeny.[21] With sigma virus a distinction exists between the capacity to initiate infection of the germinal tissue and the capacity to replicate in it; some mutants called g⁻ are transmitted hereditarily, hence they are able to replicate in germ cells, but following inoculation in flies, they fail to invade these cells.[11,12] The tested vesiculoviruses may have behaved in *Drosophila* like g⁻ sigma virus strains.

C. The Sensitivity to Refractory Genes

To this date five *Drosophila* genes interfering with sigma virus development have been described.[22] They are called refractory genes. Some of their alleles are restrictive for certain strains of sigma virus. No other vesiculovirus specifically sensitive to the different allelic forms of these five genes has been found, but Piry virus shows a sensitivity to another *Drosophila* gene which does not seem to affect the sigma virus strains tested.[23] The role of these genes in the physiology of *Drosophila* is still unknown; their study may help to identify some of the host factors interacting directly with the viral products.

In order to study in detail the physiology of rhabdoviruses in insects at the cellular and molecular level, *Drosophila* cell lines derived from insects of controlled genotype are a useful tool. Moreover, this type of research benefits from the considerable amount of information on the molecular biology of the vesiculoviruses accumulated over the past decade.[24]

VSV is a negative-stranded RNA virus which multiplies in the cell cytoplasm. All of its genetic information consists of a single chain of 11,162 nucleotides coding for five proteins, the large protein L, the glycosylated protein G, the nucleoprotein N, the membrane matrix protein M, and a phosphorylated NS protein. About 2000 molecules of N interact with the genome to form a helical ribonucleoprotein (RNP) which serves as the template for transcription by a polymerase packaged in the virion. There are two proteins, L and NS, closely associated to the RNP and responsible for this activity, L being the polymerase and NS a phosphorylated ligand indispensable to the elongation of the nucleotide chains initiated by the polymerase.[25,26]

The virion assembles at the plasma membrane and is released from the cell by budding. A lipid bilayer derived from the plasma membrane envelops the virus and contains the viral G protein, forming spikes at the virion surface. G is synthesized on membrane bound polysomes. The nascent chain is transported across the rough endoplasmic reticulum, glycosylated during the transfer, and remains anchored in the membrane by its COOH terminal where assembly and budding of the virions takes place.[27,28] The inner side of the envelope includes the M protein, which among other functions acts as a transporter of the RNP from the cytoplasm to the assembly site.[28]

The strategy of nucleic acid synthesis includes two distinct processes, the transcription of the genome into several monocistronic mRNA, and the replication of the genome via an equal size plus-stranded RNA.[25,29] Viral transcription can be carried out in vitro by the virion polymerase in the absence of cellular constituents. Accordingly, this primary transcription can be observed in cells in the presence of protein synthesis inhibitors. It results in six RNA species from the 3′ end of the genome, a 47 nucleotide sequence which is the leader RNA, and five mRNA which are capped, methylated, and polyadenylated. This transcription is sequential and proceeds in the following gene order: 3′ — leader — N — NS — M — G — L — 5′. The leader gene contains a promoter sequence that is recognized by the RNA polymerase and also the complementary sequence of a nucleation site for the assembly of nucleocapsids.[30] Symmetrically, a 59 nucleotide noncoding sequence follows the L gene at the 5′ end of the genome; it contains a nucleation site and the complementary sequence of the promoter required for genome replication.[31] Along the genome, short and highly conserved intergenic sequences signal the termination of the mRNA, the polyadenylation, and the initiation of the next mRNA.[32] At each of these signals, the polymerase can either detach itself, proceed with transcription, or initiate on internal position.[33] Different host and viral factors may control these alternatives. They have not yet been identified, but the presence of cell extracts in in vitro transcription cocktails greatly enhances the synthesis of proper mRNA transcripts.[34,35] Moreover, viral mutants with a temperature-sensitive, host-dependent, transcriptive activity have been described.[36]

Following primary transcription, viral proteins are synthesized and begin to interact with the polymerase products. This step is essential for switching the enzyme activity from transcription to replication in which full size antigenomes of plus sense are synthesized and serve, in turn, as templates for progeny negative-stranded genomes. Such products are never found in infected cells as free RNA; the nascent chains are immediately complexed with the N protein. A model has been proposed in which the increasing intracellular level of the N protein determines this modification of the polymerase activity by switching off the attenuation signal at the end of the leader sequence.[31,37] The M protein which may be phosphorylated at different degrees by kinases from the host[38,39] has also been shown to intervene in the transcriptive attenuation process.[40] Once progeny RNP have been produced, some of them bud out into viral particles while others enter a process of secondary transcription amplifying viral synthesis and so on, as long as the host cell permits it.

Defective genomes with sequence rearrangements are frequently observed in rhabdoviruses. Several models which attribute their origin to replication accidents have been proposed.[41] They consist either of simple deletions of internal sequences or deletions associated with substitutions of the 3′ end of the minus strand by 3′ end of the plus strand. These structures mature as defective interfering particles (DIP) and compete with the normal virus in mixed infections. They are unable to replicate RNA without an infectious helper virus. Some host factors are responsible for the generation of these DI particles; in human cells, chromosome 16 was shown to code for one of them.[42]

In vertebrate cells, the cytopathic effect of VSV is immediate and results in a rapid inhibition of host RNA, DNA, and protein syntheses. The viral structural constituents are not toxic per se, since infection either by UV-irradiated virions or by some types of DI particles devoid of transcriptive activity is not lethal to the cells, even at very high multiplicity. The cell killing effect can be produced by uninfectious viral particles unable to achieve a complete replication cycle. It was demonstrated in many different ways that the only requisite event was the primary transcription of the leader sequence.[43] McGowan et al.[44] showed that the cell RNA polymerases II and III are inhibited by the addition of VSV leader RNA to in vitro specific transcription systems. Moreover, VSV leader RNA was shown to be present in the nucleus of infected cells[45] where an inhibitor of host transcription would be expected to act. The mechanism of leader RNA activity is not yet completely understood. Sequence examination revealed an AU- or TATA-like region and a purine rich sequence near its 3′ end, which resemble promoter-like elements of RNA polymerases II and III, respectively, and thus could possibly serve as inhibitors of host transcription, possibly by complexing a cofactor of these enzymes.[30,44] The inhibitory effect of VSV leader RNA was recently shown to be attributed to its primary structure itself, but also to its secondary structure.[46] However, in the cell the VSV leader RNA is associated with proteins, and whether these protein associations would enhance or diminish its role is not known. Complexes with the viral N protein exist,[31] and also with a particular cellular protein[30] (the La protein) which was found to form complexes with a heterogeneous spectrum of RNAs, mostly nuclear.[47,48] The exact cellular function of the La protein is not yet known. It could either be a cofactor for RNA polymerases II and III or act in the maturation of RNA transcripts; its involvement with the VSV leader RNA may give a clue to our understanding of the leader RNA function. Still, this hypothesis awaits further investigations.

If the leader RNA is the major factor for cell killing, what happens when the cells escape the fatal course and establish persistent infections? In fact, it clearly appears that every time persistent infections of VSV have been obtained in vertebrate cells, ample concessions were required from the infecting virus essentially concerning its transcriptive activity.[49,50] This is not necessary in insect cells. In early studies of insect cell infections by rhabdoviruses, aside from the observation that the replication of virions lethal to vertebrate cells produced no cytopathic effect, attention was focused rather on the dynamics of long term persistence[2,51,52] and on the consecutive genetic drift of the viral population[2] than on the events occurring in cells during the first replication cycle of the infecting virus. Indeed these events are decisive in the establishment of persistence which implies that the infected cell has the ability to neutralize killing factors (either by inhibiting their synthesis, or by ignoring them) and to restrict any viral synthesis within limits compatible with its own survival. Therefore, in this chapter, emphasis is placed on the host control over the viral macromolecular syntheses within the first replication cycle following single step infection. Also, we will examine how the studies on the *Drosophila* genes refractory to sigma and Piry viruses might enhance our understanding of the role of host functions in rhabdoviridae physiology.

II. RHABDOVIRUS (VSV, PIRY, AND SIGMA) MULTIPLICATION IN *DROSOPHILA* CELLS

A meaningful analysis of the processes that ultimately result in the persistent infection of a cell culture should reflect as much as possible the events occurring at the level of the single cell. Thus, all the causes of heterogeneity should be avoided. The ideal conditions would consist of a genetically homogeneous and synchronous cell population infected simultaneously by a homogeneous viral population.

The viral population is readily homogenized by cloning.[53] The immunofluorescent staining of infected cells is convenient to verify that all cells of the culture are able to produce viral antigens and to determine the input multiplicities necessary for realizing single step infections.[54] This preliminary analysis is necessary since the size in particle number of the infectious unit can vary in large proportions according to the host cell system.

The homogeneity of the cell culture can be more or less achieved by cloning. When this was done on the *Aedes albopictus* cell line of Singh, the isolated clones showed a large variety of responses to infection by Sindbis or Semliki Forest virus[55,56] and also by VSV.[52] Another possibility is to use cell cultures derived from inbred insects. This can easily be performed with *Drosophila*.

A. Characteristics of *Drosophila* Cell Lines Used

Since the first establishments of *Drosophila* cell lines,[57,58] the collection of cell lines carrying genetic markers has increased considerably,[59] but only a few of them have been used to study the replication of viruses. In our laboratory, using the method of Echalier and Ohanessian,[60] primary cultures were established from 6- to 12-hr-old *Drosophila* embryos.

Care was taken to control the genotype of the different cell lines. They were issued from fly strains rendered partially isogenic by introducing reference chromosomes, and they carried alleles of refractory genes which were either restrictive or permissive to defined sigma or Piry virus mutants. Line 75E7 was derived from flies homozygous for the recessive marker ebony and the allele *ref(3)O* which is restrictive for some sigma virus strains. Line 77OM3 was derived from the fly strain Oregon M carrying permissive alleles for almost all of the sigma virus strains at the loci of each of the five known refractory genes. Two other lines were isolated to study the effect of *ref(3)A* host gene on Piry virus multiplication. This gene carried by chromosome III is present in a permissive allelic form for the Piry mutant ts1 in the Vermilion brown *Drosophila* strain and in a restrictive form in the Paris strain.[23] The line 81VVV was derived from the Vermilion brown strain and the line 81VVP12 from flies in which chromosome III of the Vermilion brown strain was replaced by that of the Paris strain.

The population doubling time was between 24 and 48 hr at 25°C according to the cell line. These lines exhibited the same morphology: small and round cells growing first to a confluent monolayer, then forming clumps of multilayered cells if they were not subcultured, thus showing no contact inhibition. Their karyotypes were predominantly diploid. A regular survey of their possible contamination by viruses which could establish persistent infections and impair further studies was performed. Two approaches were used, electron microscopy after negative staining of cell culture supernatants and inoculation of these supernatants into virus-free flies in order to detect any known symptom induced by viruses of diverse families (picorna, birna viruses, etc.).[11,61,62]

B. Infection Conditions

Since single step infections require high input multiplicities and because defective interfering particles (DIP) interfere with the replication of complete virions and are important factors in the establishment of rhabdovirus persistent infections in vertebrate cells, it was important to control the presence of DIP and to define precisely the infection conditions.

To eliminate DI particles, the strain of VSV Indiana was cloned twice on chicken embryo cells. Among the Piry virus strains isolated by Brun,[23] the two strains studied in *Drosophila* cells generated DI particles very easily and showed a high mutability; four successive clonings were necessary and the maintenance of the virus characteristics was verified at each step (in particular, the thermosensitivity of the mutant virus strain). Then the cloned VSV and Piry viruses were used to infect vertebrate cells at low infection multiplicity (less than 0.01 PFU per cell) and gave initial virus stocks of high titer. High titering stocks of sigma virus are difficult to obtain since no cell host has been found that replicates the virus with great efficiency.[9] Therefore, they were obtained by concentration of the supernatants of infected *Drosophila* cells by high speed centrifugation.[63] The observation by electron microscopy of these three virus stocks did not reveal any significant amount of truncated particles.

Experimental conditions of infection to obtain about 99% infected *Drosophila* cells after 8 hr at 25°C have been determined for VSV by immunofluorescent staining. These conditions were identical whatever the *Drosophila* cell line used when the temperature and the ratio of the inoculum volume to the surface area were identical.[63] This permitted definition of a standard procedure for infection with VSV; cells were grown until they formed a dense monolayer (about 1.6×10^6 cells/cm²) and the inoculum (80 $\mu\ell$/cm²) contained at least 2.10^9 PFU/mℓ. Similar conditions were used with Piry virus. In a first attempt to obtain single step infection with sigma virus, infection was carried out under the conditions described above with an inoculum containing 2.7×10^9 IU/mℓ (one IU corresponds to the lowest amount of virus able to induce the CO_2 sensitivity symptom in a fly after 20 days at 25°C).[11] Increase in concentration of the inoculum did not significantly change the kinetics of virus production. However, differences exist in the ability of a virus to induce CO_2 sensitivity symptom in *Drosophila*, to infect homologous cultured cells, and to give rise to plaques of lysis in vertebrate cells. Immunofluorescent studies using antisera prepared against sigma virus are now in progress and will permit the defining of more precise conditions required for a single cycle with this virus.

C. Comparative Growth Cycles of Rhabdoviruses

Mudd et al.[2] have described the multiplication of VSV in a continuous cell line derived from Oregon R *Drosophila* embryos in which persistent noncytocidal infections were obtained. In the virus production two phases could be distinguished, a regular increase during the first 20 hr with a maximum yield per cell varying between 2×10^{-2} and 2 PFU according to the VSV strain, and then a plateau corresponding to a reduced virus yield characteristic of persistent infection in insect cells. The authors did not emphasize this phenomenon which reflects a decisive step in the establishment of persistence since they were interested mainly in the selective conditions for the viral population in insect cells.

We compared the production kinetics of three rhabdoviruses (VSV, Piry, and sigma viruses) in *Drosophila* cells infected under the conditions described in the previous section. The virus released in the culture medium during 2-hr intervals was titrated at various times after infection (Figure 1). VSV and Piry virus growth curves are very similar; after a lag period of about 3 hr, the production increased rapidly up to a maximum attained 8 hr after infection, then decreased and reached a plateau at 15 hr.

The maxima of virus production were reached sooner and were higher than those observed by Mudd et al.[2] It is difficult to analyze the factors accounting for these differences, as they may result altogether from the virus (presence of DI particles), the cells, and the conditions of infection. Nevertheless, the yields of virus were low compared to those obtained in any vertebrate cell system.

Infection of uncloned mosquito cells with rhabdo- or alphaviruses resulted in virus growth curves similar to those described here.[64] Cloned *Aedes albopictus* cells showed a broad spectrum of responses to infection; the different clones gave either high yields of virus and showed moderate to severe cytopathology, or low yields of virus and no discernible cytopathology.[55,56] About 30% of the clones belonged to the former type. The results obtained with *Drosophila* cells do not pertain to the genetic heterogeneity of the cell population. This was controlled by studying the replication of VSV in 18 clones isolated from the 77OM3 *Drosophila* cell line using the method of Richard-Molard and Ohanessian.[65] The responses to VSV infection of each clone and of the uncloned population were identical.[66] This result indicates that the cell population was genetically homogeneous and did not contain various cell types.

The inhibition of virus production after a phase of intense release was not observed when actinomycin D was added before infection. Moreover, the virus yield was consid-

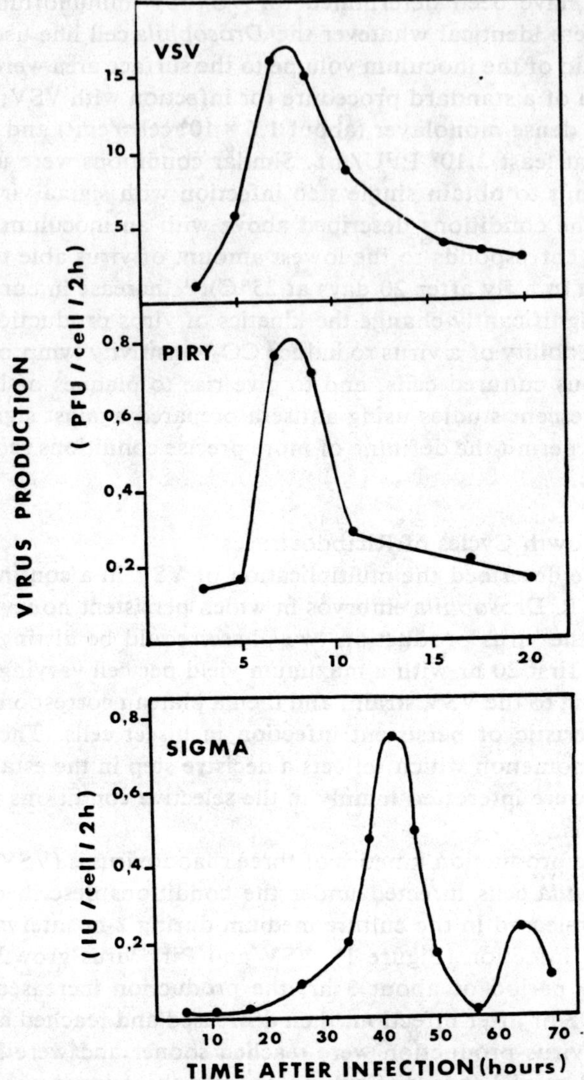

FIGURE 1.   Kinetics of VSV, Piry, and sigma virus production in *Drosophila* cells, line 75E7, 81VVV, and 77OM3, respectively. Cells were infected and processed as previously described.[3,63] Results are expressed in PFU or IU per cell per 2 hr intervals.

erably enhanced and almost reached the level observed in vertebrate cells.[3] This suggests that a cellular mechanism controls the virus multiplication. Such a control could be induced after infection since it requires a certain period of time to settle.

The sigma virus proceeded with identical steps to those of VSV or Piry virus, but the cycle developed more slowly; the lag time was more than 1 day, the production peak lasted 20 hr, and the state of persistent infection required 3 days to settle. The yield of virus per cell (defined in IU) reached a maximum value 42 hr after infection then decreased, but a second small rise often followed, suggesting that all the cells were not primarily infected. Sigma virus development was also studied in *Drosophila* flies using ts conditional-mutants.[67] The description of the biological events and the deduced timing for each step of the virus cycle were consistent with the data obtained in cultured cells. The delayed development of the sigma virus could be the expression of particular virus-host interactions resulting from the selection pressure on both the virus and its unique obligatory host. In particular, the sigma virus has to preserve its unique way of propagation (the hereditary transmission) so competition with cellular syntheses has to be reduced.[11]

III. PERSISTENT INFECTION IN INSECT CELLS

A. Differences Between the Mechanisms of Persistent Infection in Insect and Vertebrate Cells

Infection of vertebrate cells by wild-type viruses does not always result in the total destruction of the cell population. Some cells survive the primary infection and can be subcultured. They contain viral antigens, cannot be superinfected with a homologous virus unless cured cells appear, and release low amounts of viral particles in a cyclic manner. Persistently infected insect cells exhibit the same characteristics, but except for certain clones of *Aedes albopictus* cells the state of persistent infection is established immediately after the primary infection in the whole insect cell population.

Several hypotheses have been put forward to account for persistent infections with many different virus groups in vertebrate cells. The mechanisms implicated have been exhaustively reviewed.[24,49,68] They include the presence of DIP, the rapid generation of ts mutants, and the induction of antiviral substances as in the interferon system. However, the mechanisms responsible for the initial infection step are not always clearly differentiated from those which maintain the state of persistent infection.

These mechanisms are not better understood in arbovirus-infected insect cells. DI particles have been shown to be synthesized and released from *A. albopictus* cells persistently infected with the Sindbis virus for several weeks,[69] but they were not detected during the primary infection. In fact, the acute phase of Sindbis virus multiplication lasts 48 hr in mosquito cells, then the production rate declines. The cells were unable to generate DIP during the first 16 hr after infection, even after 50 serial undiluted passages using 16 hr harvests. DIP which interfere with the replication of standard Sindbis virus had already appeared after 5 passages if 48 hr harvests were used.[70] Nevertheless a particular clone of mosquito cells producing high virus yields with a drastic cytopathology rapidly generates DIP when infected by VSV or Semliki Forest virus.[52,56] However, this clone mimics a vertebrate system since only a small number of cells survives the initial infection to give a persistently infected culture.

We investigated the production of DIP by VSV-infected *Drosophila* cells. VSV yields were measured during undiluted passages at 24 hr intervals (Table 1); no drop of the viral titer was found under these conditions. DIP of VSV and Sindbis virus generated by mosquito cells has been shown to inhibit the multiplication of standard viruses only in these cells and not in vertebrate cells,[52,71] thus, interference by hypothetical DIP previously assayed in vertebrate cells[3] was also assayed in *Drosophila* cells (Table 2). Viruses were harvested during the phases of acute, declining, and low production; the release of DIP was evaluated for each of these steps, after concentration of the culture supernatant by centrifugation. In all cases, interference in *Drosophila* cells was very weak, as previously observed with vertebrate cells.[3] It increased slightly in the presence of the late harvest (VSV$_{D24-48}$) (Table 2) which suggested that a few DIP could be released late after infection. Therefore it seems that a rapid and important production of DIP cannot explain the control of VSV multiplication by *Drosophila* cells.

In vertebrate cells, some ts virus mutants could induce persistent infection at the nonpermissive temperature, but in insect cells, which are only maintained at low temperatures (25 to 28°C), persistent infection could not be due to ts mutants which only

Table 1
VSV YIELD DURING FOUR UNDILUTED PASSAGES IN DROSOPHILA CELLS

	Passage	Titer of the culture Supernatant (PFU/ml)[a]
	1	2.3×10^8
	2	3.4×10^8
	3	2.6×10^8
	4	1.5×10^8

[a] *Drosophila* cells were infected with VSV (2×10^9 PFU/ml); 1 hr later the inoculum was removed, cells were washed twice, and 2 ml of fresh medium were added. The culture supernatant was collected after 24 hr at 25°C (passage 1) and used undiluted to infect *Drosophila* cells for passage 2. This schedule was repeated for the successive passages. Virus titers were determined by plaque assay in chicken embryo cells.

Table 2
TEST FOR PRESENCE OF DI PARTICLES DURING VSV CYCLE IN DROSOPHILA CELLS BY INTERFERENCE IN DROSOPHILA CELLS

Infection schedule[a]		Virus yield at 24 hr (PFU/ml)	Percentage of VSV wt yield
−2 hr	0 time		
0	VSV wt	8×10^7	100
VSV$_{D1-8}$	0	2.45×10^7	
VSV$_{D1-8}$	VSV wt	7×10^7	87
VSV$_{D8-24}$	0	3.2×10^7	
VSV$_{D8-24}$	VSV wt	6.3×10^7	79
VSV$_{D24-48}$	0	2.7×10^7	
VSV$_{D24-48}$	VSV wt	4.9×10^7	61

[a] *Drosophila* cells were infected with VSV as described in Table 1. The virus released from 1 hr to 8 hr (VSV$_{D1-8}$), from 8 hr to 24 hr (VSV$_{D8-24}$) and from 24 hr to 48 hr (VSV$_{D24-48}$) was concentrated by high speed centrifugation, resuspended in culture medium (2×10^9 PFU/ml), and used to infect *Drosophila* cells. Two hours later (0 time) the cells were washed and superinfected with cloned VSV (wild type, wt) at 2×10^9 PFU/ml or mock-infected. Incubation at 25°C was continued for 24 hr and virus yield determined by plaque assay on chicken embryo cells.

appear randomly in the absence of selection pressures during long-term infection of cultured cells or *Drosophila* flies.[2,64,72]

No interferon-like substances have so far been found in insect cells. Several reports have described the properties of an antiviral agent released by mosquito cells persistently infected with alphaviruses.[73-75] Its effect is prevented by pretreating the cells with actinomycin D and the agent is destroyed by heat or proteinase K. The treatment of uninfected cells by this agent inhibits the RNA and protein syntheses of homologous virus. A heterologous inhibition strictly limited to the alphaviridae has also been observed, but all reports do not agree on this point. No such antiviral activity could be demonstrated in mosquito cells infected by flavi, bunyaviruses, or VSV. This activity seems specific of alphaviruses and has no significance in the establishment of persistent infection by other arboviruses. The agent is not active in vertebrate cells and, to our knowledge, it has not been tested in other insect cell species.

B. Compared Synthesis of Viral Components in *Drosophila* and Vertebrate Cells

Drosophila cell clones in which the virus could develop a lytic cycle were not found, so we used vertebrate cells as a reference system for lytic conditions. Because in *Drosophila* cells the expression of the three apparent phases of VSV multiplication was achieved within less than 15 hr, we studied the viral syntheses during this period. No inhibition of cellular protein and RNA syntheses occurred even transiently; however, the use of inhibitors to diminish the cellular synthesis background such as actinomycin D was excluded since this drug was shown to disturb the interactions between the virus and the host cell.[3]

1. VSV Protein Synthesis

In vertebrate cells each of the five viral proteins is easily detected after pulse labeling since the host macromolecular syntheses are rapidly inhibited, but among the *Drosophila* cell proteins, only four viral proteins, N, NS, M, and G were found.[3] The viral protein L, implicated in viral RNA syntheses, was not detectable; it may have been present in too small a quantity or hidden by cellular proteins.

The kinetic studies (Figure 2) showed that in *Drosophila* cells the synthesis rate of each viral protein reached a maximum between 5 and 7 hr after infection, then dropped to a constant and low level. Moreover, the glycoprotein G, the viral outer membrane protein, and its shorter form G1 were always poorly synthesized. Thus, the G + G1 synthesis rate relative to that of the N protein was five-fold lower in *Drosophila* cells than in chicken embryo fibroblasts (CEF) 5 hr after infection. G protein deficiency was also observed in the mature virions released from *Drosophila* cells (Figure 3.1).[3] The synthesis of each viral protein was increased upon actinomycin D treatment just before infection, and the most stimulated protein was the G. In spite of this, the mature virions still contained as little G protein as in the absence of the drug.[3]

Several hypotheses may explain these results: (1) G protein is glycosylated and sialylated by cellular enzymes; it has been shown that sialic acid is lacking in all viruses grown in insect cells as in the host cell itself.[64] This deficiency which is also observed after actinomycin D treatment, could impair G protein incorporation in the cell membrane and in the virus envelope; (2) G protein is known to be the only viral protein translated on membrane-bound polyribosomes where a competition between viral and host protein translation could be more severe towards the virus. This competition could favor G protein synthesis in the presence of actinomycin D; and (3) G mRNA and L mRNA are, respectively, the penultimate and the ultimate viral mRNA synthesized during the sequential transcription of VSV genome. A drastic attenuation at the level of viral transcription could result in a G and L protein deficiency. To determine

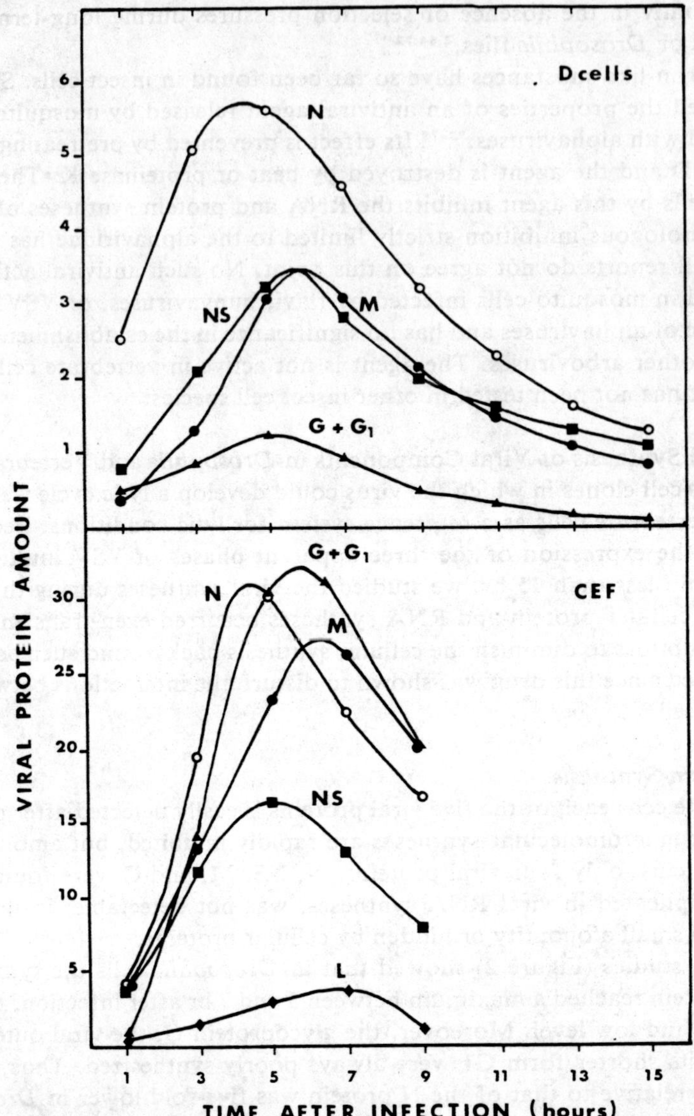

FIGURE 2.   Kinetics of VSV polypeptide synthesis in *Drosophila* and chicken embryo cells. At various intervals after infection with VSV the cells were pulse labeled with ^{14}C-amino acids for 30 min. Proteins were separated by electrophoresis on a polyacrylamide slab gel. The gel autoradiography was scanned and the amount of each viral polypeptide was calculated as previously described.[3]

whether transcription or translation of G mRNA is impaired in *Drosophila* cells, the molar ratios of the various mRNAs have to be determined.

The low synthesis rate of the G protein and its weak incorporation into the virions have also been observed in a particular clone of mosquito cells capable of producing high yields of VSV with a cytopathic effect as important as in vertebrate cells.[76] Therefore these two singularities do not seem to be directly associated with the development of persistent infection in insect cells.

The inner-membrane M protein of VSV is a phosphoprotein and its phosphorylation depends on host cell kinases.[77] It was found three and a half-fold more phosphorylated

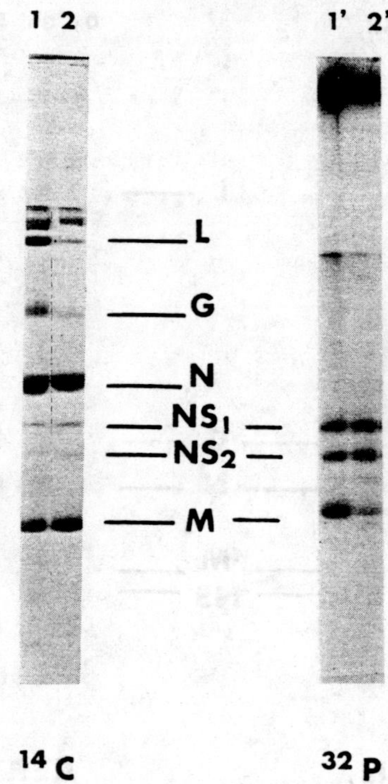

FIGURE 3.   Proteins and phosphoproteins of mature VSV virions grown in *Drosophila* or chicken embryo cells. *Drosophila* cells (1 and 1′) or chicken embryo cells (2 and 2′) were infected with VSV and labeled with either (^{14}C)-leucine or [32]Pi.[3,78] The proteins of the purified virus were separated on a polyacrylamide slab gel containing 7M-urea and detected by autoradiography.

in the mature virions released from *Drosophila* cells than in those from CEF cells (Figure 3.1′). This difference was detected in the cell cytoplasm as soon as the protein was synthesized.[78] On the contrary the other viral phosphoprotein NS (which can be separated into different phosphorylated forms [NS1 and NS2]) did not differ.[77,78] Partial proteolysis of ^{32}P-labeled M protein synthesized in the two cell types allowed the comparing of the phosphorylated peptides. One of them which contains at least one site of phosphorylation in CEF was found devoid of phosphate in *Drosophila* cells except after actinomycin D treatment. All others were phosphorylated, in both actinomycin D treated and untreated *Drosophila* cells, but at a higher degree than in CEF.[78]

The M protein takes an important part in multiple viral functions; it acts as a regulator in the RNA transcription-replication steps and plays a pivotal role in the budding process by transporting the nucleocapsid to the cellular membrane and by interacting with the G protein.[77,79,80] It is not known how much the phosphorylation level of each site and their selection can affect these functions and whether, for instance, they could explain the protein G deficiency in mature virions, but such host effects are important and should be kept in mind.

Modification and/or restriction of the expression of the viral proteins associated with the membrane have often been observed in vertebrate cells, either persistently infected or in an antiviral state after interferon treatment. In the former case, mutated forms of these proteins were found in VSV, rabies, and Sindbis virus;[81-83] the amount of Sendai virus M protein was reduced because of its instability;[84] and the phosphoryl-

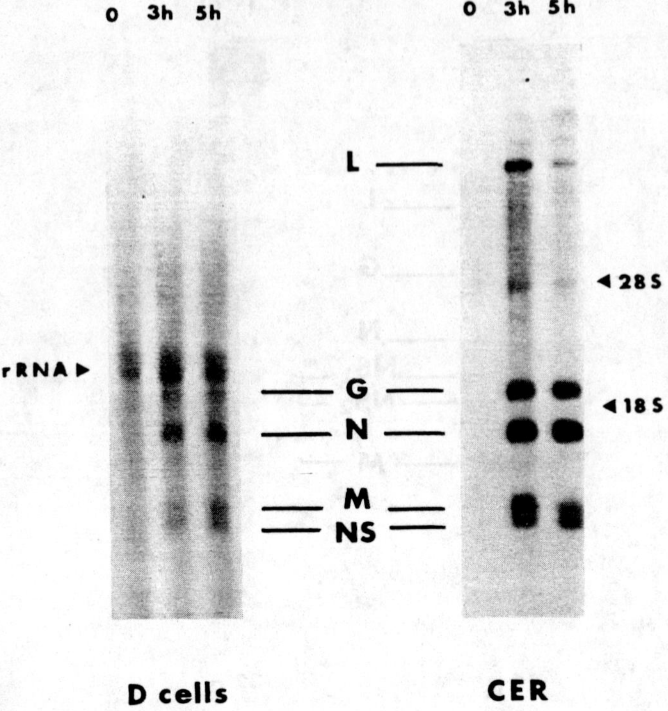

FIGURE 4.   Detection of the VSV mRNAs present in infected *Droso-phila* and CER cells. Intracellular RNAs from *Drosophila* or CER cells were extracted 0, 3, or 5 hr after infection with VSV; aliquots correspond-ing to 10^6 and 10^5 cells, respectively, were separated on an agarose-urea gel,[37] transferred on a gene screen membrane, and hybridized with VSV genome RNA, ^{32}P labeled by polynucleotide kinase. The position of the ribosomal RNAs from *Drosophila* and CER cells is indicated.

ation of rabies virus M_1 protein was enhanced, while the synthesis of G and M_2 was ten-fold lower than in acutely infected cells.[85] Vertebrate cells treated by interferon and infected with VSV also produced viral particles containing low amounts of G and M proteins, probably due to the inhibition of membrane protein incorporation during the virus assembly.[86] In all cases, the cellular control over the virus multiplication has not been clearly correlated either with quantitative and qualitative modifications of these proteins, or with an alteration of their functions.

2. VSV RNA Synthesis

We have studied VSV RNA synthesis in cultured *Drosophila* cells using different methods of hybridization which allow the detection of specific RNAs among the cel-lular nucleic acids and the quantifying of them, even if they are present in small amounts.

a. VSV Transcription in Drosophila Cells

At different times after *Drosophila* cell infection, a radiolabeled viral genome probe was hybridized in liquid phase to the whole intracellular RNAs. The viral positive-stranded RNAs recovered consisted mainly in mRNAs as shown below and in Figure 4. Their total amount first increased until 7 hr, then rapidly decreased to a lower level representing, 11 hr after infection, about 20% of the maximum.

Each of the different viral mRNAs was analyzed separately after electrophoresis, transfer, and hybridization to a radiolabeled genome probe (Figure 4).[37,87] This method permits detection in vertebrate cells of the five viral mRNAs as well as the antigenome. In *Drosophila* cells, only the four mRNAs corresponding to G, N, NS, and M proteins appeared as early as 3 hr after infection. The L mRNA was not abundant enough to be detected in this experiment. The ratios of each one of the other mRNAs relative to the N protein mRNA were calculated after scanning the autoradiograms; they were identical in the two cell types. Thus, in *Drosophila* cells, the synthesis rate of the viral N, NS, and M proteins appeared proportional to the amount of message, while the G protein deficiency seemed to result from an inhibition at the translation level. Still, all along the virus cycle in *Drosophila* cells the total amount of viral mRNAs always remained much lower than in vertebrate cells suggesting either a lower synthesis rate or a rapid degradation. The kinetics of viral mRNA synthesis were studied using a dot-blot-hybridization technique.[88] Intracellular RNAs were pulse-labeled in vivo and viral positive-stranded RNAs specifically hybridized to the unlabeled immobilized viral genome. In vertebrate cells, the rate of VSV RNA synthesis increased up to 3 hr and remained constant until cellular lysis. In *Drosophila* cells, it also increased during the first 3 hr, but dropped down very rapidly and was already minimal 7 hr after infection (data not shown). Thus, a particular degradation mechanism of VSV mRNAs in *Drosophila* cells is unlikely since messengers are still present and translated several hours after transcription inhibition. On the contrary, these results suggest that a cellular control may act on VSV RNA synthesis itself or on a former step of the virus development.

In the same way, Tooker and Kennedy[56] have shown that in a low virus-producing line of mosquito cells, restriction of Semliki Forest virus growth arose at the level of negative-stranded RNA synthesis. Recently, short viral RNAs were detected within the cells, but these RNAs were not released from the cells as defective viral particles.[89] A similar situation has also been described for vertebrate cells persistently infected with Sindbis virus.[83] The study of negative-stranded viral RNA synthesis within *Drosophila* cells is now necessary to determine if intracellular truncated RNAs are synthesized and could be implicated in the establishment of persistence by interfering with VSV transcription or replication.

b. Synthesis and Localization of VSV Leader RNA in Drosophila Cells

The role of VSV leader RNA in the shut off of vertebrate cell transcription has been clearly demonstrated. The lack of such an effect in *Drosophila* cells could be explained if the synthesis of VSV leader proceeded differently, if it was not transported into the cell nucleus, or if it had no effect on the transcription by the insect RNA polymerases.

Thus, a study of the VSV leader RNA in *Drosophila* cells was initiated. This RNA was specifically detected by hybridization with the viral genome 3′-end labeled by T4 RNA ligase.[90] After RNase digestion of single-stranded RNA, double-stranded hybrids were analyzed by electrophoresis on a polyacrylamide gel. Two short RNA molecules both corresponding to the leader RNAs described in vertebrate cells were present in the infected *Drosophila* cells (Figure 5). Preliminary results indicated that after cell fractionation, leader RNA was found in the cytoplasm, and small amounts could be also detected in the nucleus. Thus, VSV leader RNA was synthesized in *Drosophila* cells and seemed to migrate into the nucleus. Nevertheless, its amount was always much lower than in vertebrate cells where more than 1000 leader copies were found per cell.[30,45] If a threshold level of leader RNA is required for host synthesis shutoff, the low amount found in the infected *Drosophila* cell nucleus could explain the absence of cytopathogenicity of VSV in these cells.

The precise mechanisms of the inhibition of cellular transcription are still unknown, nevertheless, VSV leader RNA was found to inhibit vertebrate RNA polymerase II and III transcription in vitro.[44,46] It is possible that VSV leader RNA could have no effect

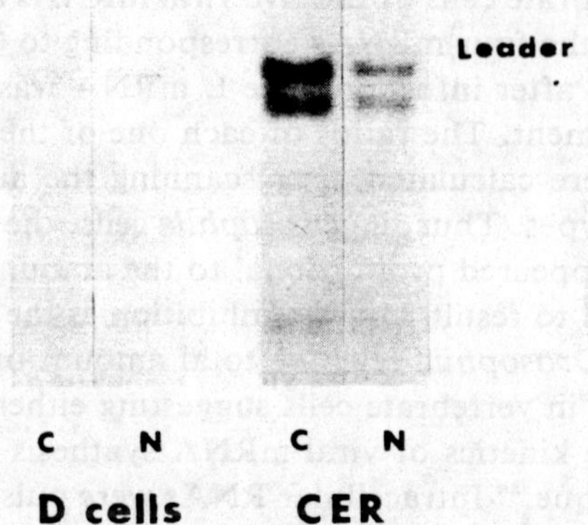

FIGURE 5.   Detection of VSV leader RNA present in the nucleus and the cytoplasm of infected *Drosophila* or CER cells. The cells were harvested 3 hr after infection with VSV. After cell fractionation, RNAs were extracted and annealed with VSV genome RNA, ^{32}P labeled at its 3′ end by T4 RNA ligase. Double-stranded RNAs resistant to RNase A + T$_1$, were separated by electrophoresis on a 10% acrylamide slab gel. The cytoplasmic fractions of both cell types corresponded to 3×10^5 cells. The nuclear fractions corresponded to 3×10^5 and 1.2×10^6 cells for CER and *Drosophila* cells, respectively. Leader RNAs were detected by autoradiography.

on insect cell transcription because of fundamental differences between the RNA polymerases of the two host cell systems.

IV. *DROSOPHILA* GENES WHICH INDUCE MODIFICATIONS IN THE MULTIPLICATION OF SIGMA AND PIRY VIRUSES: REFRACTORY GENES

A. Historical

To the present time, five *Drosophila* genes are known to act specifically in the cycle of sigma virus, *ref(1)H*, *ref(2)M*, *ref(2)P*, *ref(3)O*, and *ref(3)D*,[22] and one in the cycle of Piry virus, *ref(3)A*[23] (the number in parentheses refers to the *Drosophila* chromosome carrying this gene). A refractory gene is defined by the existence of at least two alleles, one permissive and the other restrictive for a given strain of virus. The following observations should be added to this simple definition: (1) the flies carrying either restrictive or permissive alleles express no obvious morphological or physiological differences; in particular no modifications of their lifespan or fertility could be observed; (2) for each gene it has been possible to select a virus strain for which an allele, defined as restrictive toward other strains, is permissive; and (3) all the allelic forms recognized for the refractory genes preexisted in nature or in laboratory strains.

Biological methods permitted definition of the cycle of the sigma virus in flies. Following infection of flies by injection, the first step is the adsorption-penetration process (step Ao in Figure 6). This step was fast since inoculation and reextraction of infectious units (IU) showed that 99% of them disappeared within 1 hr.[91] Immediately thereafter, the decay of infectious centers at high temperatures could be observed. It followed typical single hit inactivation kinetics and the destroyed unit appeared to be the viral genome. Thus, it was possible to distinguish between an infected cell containing a single viral genome and that containing several and, therefrom, to locate in time the viral genome replication along the virus cycle.[67,91] It should be noted that such experiments are meaningful only because the sigma virus genome shows a high ther-

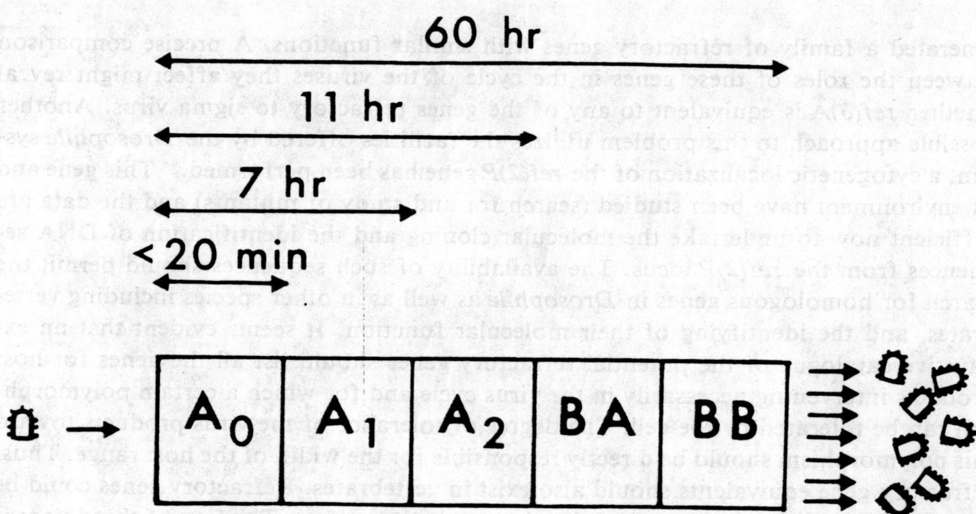

FIGURE 6. Formal description of sigma virus cycle in *Drosophila* flies under standard conditions (Oregon × ebony heterozygous host, incubation at 20°C). The mean time necessary to perform each step is shown. The steps named A are those for which only one viral genome per cell is detectable. The steps named B are those for which several genomes per cell are present, e.g., after the viral genome replication or during the carrier state. The step named BB is the viral maturation. Only events occurring during the BA step are necessary to perpetuate the carrier state.

molability. The genome of vertebrate rhabdoviruses is generally much more stable, except for some mutants like the VSV mutant tsO52.[92]

For ts mutants of sigma virus, the study of the infectious center decay and of the temperature sensitive period allowed recognition of two types of mutants affected prior or during the viral genome replication, and so to define two successive steps in the viral cycle, A_1 and A_2 (Figure 6).[91] After the A steps, the infected cell contains several viral genomes. In this state, the infectious centers can still be destroyed by heat treatment, but their number decreases 5 to 10 times more slowly than during the A steps.[67,91] This second phase of inactivation can be observed only when the infection of neighboring cells is impeded, i.e., when viral maturations are blocked either directly in the case of ts mutants for maturation functions, or indirectly if the secondary transcription is inhibited (B steps). Simple tests concerning the hereditary transmission at high temperatures allows distinction between the two possibilities. The timing observed for the different steps defined along the first viral cycle following injection is shown in Figure 6.

A sigma virus mutant more ts in the restrictive host than in the permissive one was found for each refractory gene.[93] In this case, using the same biological tests, at least part of the restrictive effects of the host could be studied. For instance, the major effect detected for a restrictive allele of the *ref(2)P* gene resulted from a reduced efficiency of the A_1 step.

The prospects of such a study would be short if the functions of these genes pertained strictly to the special relation of the sigma virus with its natural host, but the Piry virus, which has a wider host range including vertebrates, is also sensitive to *Drosophila* refractory genes.[23] The *ref(3)A* and possibly two other genes which have not yet been precisely localized .were found to interact with the Piry virus. *Ref(3)A* is distinct from the refractory genes defined for sigma virus, but it is not known if it functions in a similar fashion. In the evolution of eukaryotic organisms, gene duplication followed by specialization of the homologous units according to the differentiated cell type or to the stage of development occurred frequently. Such a mechanism could have generated a family of refractory genes with similar functions. A precise comparison between the roles of these genes in the cycle of the viruses they affect might reveal whether *ref(3)A* is equivalent to any of the genes refractory to sigma virus. Another possible approach to this problem utilizes the facilities offered by the *Drosophila* system, a cytogenetic localization of the *ref(2)P* gene has been performed.[94] This gene and its environment have been studied (search for and study of mutants) and the data are sufficient now to undertake the molecular cloning and the identification of DNA sequences from the *ref(2)P* locus. The availability of such sequences should permit the search for homologous genes in *Drosophila* as well as in other species including vertebrates, and the identifying of their molecular function. It seems evident that an exhaustive catalogue of the potential refractory genes should list all the genes for host products intervening necessarily in the virus cycle and for which a certain polymorphism can be tolerated by the cell. The degree of tolerance of the virus products toward this polymorphism should be directly responsible for the width of the host range. Thus, refractory gene equivalents should also exist in vertebrates. Refractory genes could be classified according to the steps of the virus cycle they affect. The first of these steps is the entry of the virus into the cell. The methods of analysis described above revealed that the major effect of the five refractory genes could not result from a modification of the penetration efficiency of sigma virus.

B. Expression in Continuous Cell Lines In Vitro

Richard-Molard[95] has shown that the restrictive effect of the allele P^p of the *ref(2)P* gene was also expressed in two continuous cell lines homozygous for this allele. As observed in homologous flies, this effect was not absolute and a residual multiplication of the sensitive sigma virus strains was obtained. This could result from a heterogeneity of the virus population or, more probably, from a leakiness since residual multiplication was also observed in restrictive flies inoculated with cloned sensitive viruses.[96] Unfortunately this study was halted and all these cell lines discarded following their invasion by double-stranded RNA viruses of unknown origin, perhaps via calf serum.[61,62]

More recently, two cell lines, VVV and VVP, were established. They originate from flies differing only by their chromosome III on which the alleles of the *ref(3)A* gene were, respectively, permissive and restrictive toward a Piry virus mutant named ts1. The kinetics of virus production for the wild type F_7 and the mutant ts1 have been explored in both cell lines. Figure 7 shows that the multiplication of the F_7 virus was slightly reduced in the VVV cell line compared to VVP. The ts1 mutant was able to multiply in the VVV host cells as well as the F_7 virus. In the VVP cell line, ts1 multiplication was clearly impaired compared to the wild-type virus F_7 and only 2 to 3 times lower than in the permissive VVV cell line. This residual multiplication in the VVP cells was clearly higher than the growth of sensitive sigma strains in restrictive cell lines. This could neither be attributed to virus heterogeneity nor to a high frequency of reversion; on the one hand the virus was plaque-purified as described in Section 2 and on the other hand infection of VVP cells with the wild-type virus at a concentration of 1×10^6 PFU/mℓ (corresponding to a reversion frequency of 2 to 3/1000) resulted in a multiplication about 100 times lower than that obtained with the ts1 mutant. Still this result could be due to the mutual aid of the virus particles, since the multiplicity of infection was higher than one infectious unit per cell. These conditions of infection differ completely from those used when flies are inoculated. Another possible explanation for the small difference between the ts1 multiplication in restrictive and permissive cell lines could be that a set of refractory genes was implicated in the viral function impaired in the presence of the restrictive allele of *ref(3)A*; permissive forms of some of these genes could counterbalance the *ref(3)A* effect; and among the several potential

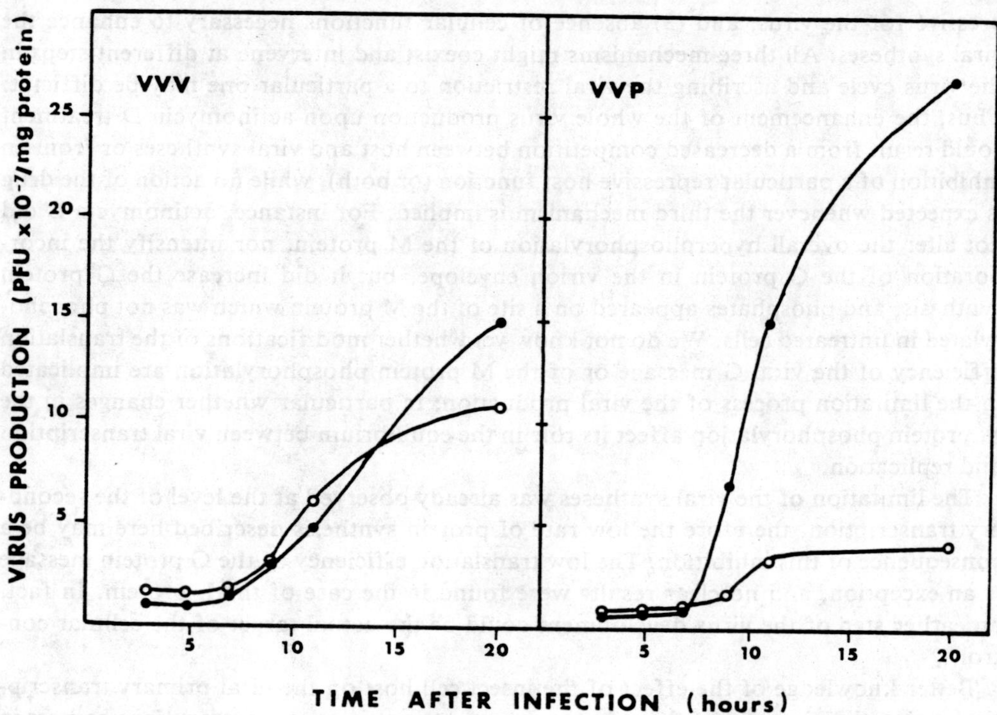

FIGURE 7.   Growth of Piry virus, wild type, F₇, (O—O), and mutant, ts1 (O—O) in the *Drosophila* cell line 81 VVV and 81 VVP, respectively. Each point represents the titration of the viruses released in the supernatant fluid at various times after infection. The virus titers were expressed in PFU per mg protein of producing cells to eliminate differences in cell numbers.

refractory genes, those expressed in a particular cell line could be different from those expressed in the target tissues implicated in the expression of the characteristics used to demonstrate the restrictive effect in flies.

Nevertheless, the impaired multiplication of the mutant Piry virus ts1 in the VVP cell line may be a consequence of the expression of the *ref(3)A* gene in cultured cells.

V. CONCLUSION

There are two aspects of persistent infection in insect cells that can be distinguished and more or less analyzed separately, the limitation of viral syntheses and the absence of host synthesis shutoff. In these two phenomena, the distinction between cause and consequence is not easy. For instance, we still ignore whether the VSV leader RNA which likely takes a prominent part in vertebrate cell killing, does not assume the same role in insect cells either because sufficient amounts cannot be synthesized or because it has no inhibitory effect on polymerases.

The absence of host synthesis shutoff somehow facilitates the study of the mechanisms which limit the viral syntheses. Mutations of the host genes modifying the efficiency of these mechanisms would be an interesting tool in the study of the functions involved. One or more *Drosophila* refractory genes might be implicated in some of these functions even in their permissive forms since the limitation observed in insect cells was always increased in the presence of the restrictive alleles.

Roughly, three kinds of mechanisms could account for the low viral syntheses: (1) competition for the use of cell functions in which the viral products are underprivileged; (2) existence of host functions (either induced or constitutive) specifically repressive for the virus; and (3) absence of cellular functions necessary to enhance the viral syntheses. All three mechanisms might coexist and intervene at different steps in the virus cycle and ascribing the viral restriction to a particular one may be difficult. Thus, the enhancement of the whole virus production upon actinomycin D treatment could result from a decreased competition between host and viral syntheses or from an inhibition of a particular repressive host function (or both), while no action of the drug is expected whenever the third mechanism is implied. For instance, actinomycin D did not alter the overall hyperphosphorylation of the M protein, nor intensify the incorporation of the G protein in the virion envelope, but it did increase the G protein synthesis, and phosphates appeared on a site of the M protein which was not phosphorylated in untreated cells. We do not know yet whether modifications of the translation efficiency of the viral G message or of the M protein phosphorylation are implicated in the limitation process of the viral production; in particular whether changes in the M protein phosphorylation affect its role in the equilibrium between viral transcription and replication.

The limitation of the viral syntheses was already observed at the level of the secondary transcription, therefore the low rate of protein synthesis described here may be a consequence of this inhibition. The low translation efficiency of the G protein message is an exception, and no clear results were found in the case of the L protein. In fact, an earlier step of the virus development could be the actual target of the cellular control.

Better knowledge of the effect of the insect cell host on the viral primary transcription and replication and of the refractory gene actions is now necessary for progress in this study.

REFERENCES

1. Singh, K. R. P., Growth of Arbovirus in Arthropod tissue culture, in *Advances in Virus Research*, Vol. 17, Smith, K. M., Lauffer, M. A., and Bang, F. B., Eds., Academic Press, New York, 1972, 187.
2. Mudd, J. A., Leavitt, R. W., Kingsbury, D. T., and Holland, J. J., Natural selection of mutants of vesicular stomatitis virus by cultured cells of *Drosophila melanogaster, J. Gen. Virol.,* 20, 341, 1973.
3. Wyers, F., Richard-Molard, C., Blondel, D., and Dézelée, S., Vesicular stomatitis virus growth in *Drosophila melanogaster* cells: G protein deficiency, *J. Virol.,* 33, 411, 1980.
4. Bras-Herreng, F., Multiplication du virus sindbis dans des cellules de Drosophile cultivées *in vitro, Arch. Virol.,* 48, 121, 1975.
5. L'Heritier, Ph., The hereditary virus of *Drosophila,* in *Advances in Virus Research,* Vol. 5, Smith, K. M. and Lauffer, M. A., Eds., Academic Press, New York, 1958, 195.
6. Berkaloff, A., Bregliano, J. C., and Ohanessian, A., Mise en évidence de virions dans des Drosophiles infectées par le virus héréditaire sigma, *C. R. Acad. Sci. Paris,* 260, 5956, 1965.
7. Matthews, R. E. F., Classification and nomenclature of viruses, *Intervirology,* 17, 109, 1982.
8. Rosen, L., Carbon dioxide sensitivity in mosquitoes infected with sigma, vesicular stomatitis and other Rhabdoviruses, *Science,* 207, 989, 1980.
9. Ohanessian, A., Personal communication.
10. L'Heritier, Ph., *Drosophila* viruses and their role as evolutionary factors, in *Evolutionary Biology,* Vol. 4, Dobzhansky, Th., Hecht, M. K., and Steere, W. C., Eds., Meredith, New York, 1970, chap. 7.
11. Brun, G. and Plus, N., The viruses of *Drosophila,* in the *Genetics and Biology of Drosophila,* Vol. 2d, Ashburner, M. and Wright, T. R. F., Eds., Academic Press, London, 1980, 625.
12. Teninges, D., Contamine, D., and Brun, G., *Drosophila* sigma virus, in *Rhabdoviruses,* Vol. 3, Bishop, D. H. L., Ed., CRC Press, Boca Raton, Fla., 1980, chap. 6.
13. Bussereau, F., Etude du symptôme de la sensibilité au CO_2 produit par le virus sigma de la Drosophile. II. Evolution comparée du rendement des centres nerveux et des divers organes après inoculation dans l'abdomen et dans le thorax, *Ann. Inst. Pasteur Paris,* 118, 626, 1970.

14. **Plus, N.**, Etude de la multiplication du virus de la sensibilité au gaz carbonique chez la Drosophile, *Bull. Biol. Fr. Belg.*, 88, 248, 1954.

15. **Printz, P.**, Relationship of sigma virus to vesicular stomatitis virus, *Advances in Virus Research*, Vol. 18, Lauffer, M. A., Bang, F. B., Maramorosch, K., and Smith, K. M., Eds., Academic Press, New York, 1973, 143.

16. **Bussereau, F., de Kinkelin, P., and Le Berre, M.**, Infectivity of fish Rhabdoviruses for *Drosophila melanogaster*, *Ann. Microbiol. Inst. Pasteur*, 126A, 389, 1975.

17. **Bussereau, F.**, The CO₂ sensitivity induced by two Rhabdoviruses, piry and chandipura, in *Drosophila melanogaster*, *Ann. Microbiol. Inst. Pasteur*, 126B, 389, 1975.

18. **L'Heritier, Ph. and Tessier, G.**, Un mécanisme héréditaire aberrant chez la drosophile, *C. R. Acad. Sci. Paris*, 206, 1193, 1938.

19. **Contamine, D.**, Etude de mutants thermosensibles du virus sigma, *Mol. Gen. Genet.*, 124, 233, 1973.

20. **Brun, G.**, Etude d'une Association du Virus sigma et de son Hôte la Drosophile: l' Etat stabilisé, Thèse Paris-Orsay, 1963.

21. **Tesh, R. B., Chianolis, B. N., and Johnson, K. M.**, Vesicular stomatitis virus (Indiana serotype) transovarial transmission by Phlebotomine sandflies, *Science*, 175, 1477, 1972.

22. **Gay, P.**, Les gènes de la drosophile qui interviennent dans la multiplication du virus sigma, *Mol. Gen. Genet.*, 159, 269, 1978.

23. **Brun, G.**, Are the *Drosophila ref* genes for piry and sigma viruses identical?, in *The Replication of Negative Strand Viruses*, Bishop, D. H. L. and Compans, R. W., Eds., Elsevier/North Holland, Amsterdam, 1981, 921.

24. **Bishop, D. H. L., Ed.**, *Rhabdoviruses*, Vol. 1, 2, and 3, CRC Press, Boca Raton, Fla., 1979, 1980.

25. **Emerson, S. U.**, Reconstitution studies detect a single polymerase entry site on the vesicular stomatitis virus genome, *Cell*, 31, 635, 1982.

26. **De, B. P. and Banerjee, A. K.**, Specific interactions of vesicular stomatitis virus L and NS proteins with heterologous genome ribonucleoprotein template lead to mRNA synthesis *in vitro*, *J. Virol.*, 51, 628, 1984.

27. **Etchison, J. R. and Summers, D. F.**, Structure, synthesis and function of the vesicular stomatitis virus glycoprotein, in *Rhabdoviruses*, Vol. 1, Bishop, D. H. L., Ed., CRC Press, Boca Raton, Fla., 1979, chap. 8.

28. **Morrison, T. G.**, Rhabdoviral assembly and intracellular processing of viral components, in *Rhabdoviruses*, Vol. 2, Bishop, D. H. L., Ed., CRC Press, Boca Raton, Fla., 1980, chap. 6.

29. **Ball, L. A. and Wertz, G. W.**, VSV RNA synthesis: how can you be positive?, *Cell*, 26, 143, 1981.

30. **Kurilla, M. G. and Keene, J. D.**, The leader RNA of vesicular stomatitis virus is bound by a cellular protein reactive with anti-La lupus antibodies, *Cell*, 34, 837, 1983.

31. **Blumberg, B. M., Giorgi, C., and Kolakofsky, D.**, N protein of vesicular stomatitis virus selectively encapsidates leader RNA *in vitro*, *Cell*, 32, 559, 1983.

32. **Rose, J. K.**, Complete intergenic and flanking gene sequences from the genome of vesicular stomatitis virus, *Cell*, 19, 415, 1980.

33. **Lazzarini, R. A., Chien, I., Yang, F., and Keene, J. D.**, The metabolic fate of independently initiated VSV mRNA transcripts, *J. Gen. Virol.*, 58, 429, 1982.

34. **Both, G. W., Moyer, S. A., and Banerjee, A. K.**, Translation and identification of the mRNA species synthesized *in vitro* by virion-associated RNA polymerase of vesicular stomatitis virus, *Proc. Natl. Acad. Sci. U.S.A.*, 72, 274, 1975.

35. **Rose, J. K., Lodish, H. F., and Brock, M. L.**, Giant heterogeneous polyadenylic acid on vesicular stomatitis virus mRNA synthesized *in vitro* in the presence of S-adenosylhomocysteine, *J. Virol.*, 21, 683, 1977.

36. **Pringle, C. R.**, The tdCE and hrCE phenotypes: host range mutants of vesicular stomatitis virus in which polymerase function is affected, *Cell*, 15, 597, 1978.

37. **Patton, J. T., Davis, N. L., and Wertz, G. W.**, N protein alone satisfies the requirement for protein synthesis during RNA replication of vesicular stomatitis virus, *J. Virol.*, 49, 303, 1984.

38. **Imblum, R. L. and Wagner, R. R.**, Protein kinase and phosphoproteins of vesicular stomatitis virus, *J. Virol.*, 13, 113, 1974.

39. **Moyer, S. A. and Summers, D. F.**, Phosphorylation of vesicular stomatitis virus *in vitro* and *in vivo*, *J. Virol.*, 13, 455, 1974.

40. **Clinton, G. M., Little, S. P., Hagen, F. S., and Huang, A. S.**, The matrix (M) protein of vesicular stomatitis virus regulates transcription, *Cell*, 15, 1455, 1978.

41. **Lazzarini, R. A., Keene, J. D., and Schubert, M.**, The origins of defective interfering particles of the negative-strand RNA viruses, *Cell*, 26, 145, 1981.

42. **Kang, C. Y., Weide, L. G., and Tischfield, J. A.**, Suppression of vesicular stomatitis virus defective interfering particle generation by a function(s) associated with human chromosome 16, *J. Virol.*, 40, 946, 1981.

43. Marcus, Ph. I. and Sekellick, M. J., Cell killing by vesicular stomatitis virus: the prototype rhabdovirus, in *Rhabdoviruses,* Vol. 3, Bishop, D. H. L., Ed., CRC Press, Boca Raton, Fla., 1983, chap. 2.

44. McGowan, J. J., Emerson, S. U., and Wagner, R. R., The plus-strand leader RNA of VSV inhibits DNA-dependent transcription of Adenovirus and VS40 genes in a soluble whole-cell extract, *Cell,* 28, 325, 1982.

45. Kurilla, M. G., Piwnica-Worms, H., and Keene, J. D., Rapid and transient localization of the leader RNA of vesicular stomatitis virus in the nuclei of infected cells, *Proc. Natl. Acad. Sci. U.S.A.,* 79, 5240, 1982.

46. Grinnell, B. W. and Wagner, R. R., Nucleotide sequence and secondary structure of VSV leader RNA and homologous DNA involved in inhibition of DNA-dependent transcription, *Cell,* 36, 533, 1984.

47. Lerner, M. R., Andrews, N. C., Miller, G., and Steitz, J. A., Two small RNAs encoded by Epstein-Barr virus and complexed with protein are precipitated by antibodies from patients with systemic lupus erythematosus, *Proc. Natl. Acad. Sci. U.S.A.,* 78, 805, 1981.

48. Rinke, J. and Steitz, J. A., Precursor molecules of both human 5S ribosomal RNA and transferred RNAs are bound by a cellular protein reactive with anti-La lupus antibodies, *Cell,* 29, 149, 1982.

49. Holland, J. J., Kennedy, S. I. T., Semler, B. L., Jones, C. L., Roux, L., and Grabau, E. A., Defective interfering RNA viruses and the host-cell response, in *Comprehensive Virology,* Vol. 16, Fraenkel-Conrat, H. and Wagner, R. R., Eds., Plenum Press, New York, 1980, chap. 3.

50. Frey, T. K. and Youngner, J. S., Further studies of the RNA synthesis phenotype selected during persistent infection with vesicular stomatitis virus, *Virology,* 136, 211, 1984.

51. Artsob, H. and Spence, L., Persistent infection of mosquito cell lines with vesicular stomatitis virus, *Acta Virol.,* 18, 331, 1974.

52. Gillies, S. and Stollar, V., Generation of defective interfering particles of vesicular stomatitis virus in *Aedes albopictus* cells, *Virology,* 107, 497, 1980.

53. Stampfer, M., Baltimore, D., and Huang, A. S., Absence of interference during high-multiplicity infection by clonally purified vesicular stomatitis virus, *J. Virol.,* 7, 409, 1971.

54. Yang, Y. J., Stoltz, D. B., and Prevec, L., Growth of vesicular stomatitis virus in continuous culture line of *Antheraea eucalypti* moth cells, *J. Gen. Virol.,* 5, 473, 1969.

55. Sarver, N. and Stollar, V., Sindbis virus-induced cytopathic effect in clones of *Aedes albopictus* (Singh) cells, *Virology,* 80, 390, 1977.

56. Tooker, P. and Kennedy, S. I. T., Semliki forest virus multiplication in clones of *Aedes albopictus* cells, *J. Virol.,* 37, 589, 1981.

57. Echalier, G. and Ohanessian, A., Isolement, en cultures *in vitro,* de lignées cellulaires diploïdes de *Drosophila melanogaster, C. R. Acad. Sci. Paris,* 268, 1771, 1969.

58. Kakpakov, V. T., Gvosdev, V. A., Platova, T. P., and Polukarova, L. G., *In vitro* establishment of embryonic cell lines of *Drosophila melanogaster, Genetika,* 5, 67, 1969.

59. Hink, W. F., The 1979 compilation of invertebrate cell lines and culture media, in *Invertebrate Systems In Vitro,* Kurstak, E., Maramorosch, K., and Dübendorfer, A., Eds., Elsevier/North-Holland, Amsterdam, 1980, 553.

60. Echalier, G. and Ohanessian, A., *In vitro* culture of *Drosophila melanogaster* embryonic cells, *In Vitro,* 6, 162, 1970.

61. Teninges, D., Ohanessian, A., Richard-Molard, C., and Contamine, D., Isolation and biological properties of *Drosophila* X virus, *J. Gen. Virol.,* 42, 241, 1979.

62. Teninges, D., Ohanessian, A., Richard-Molard, C., and Contamine, D., Contamination and persistent infection of *Drosophila* cells by reovirus type particles, *In Vitro,* 15, 425, 1979.

63. Richard-Molard, C., Blondel, D., Wyers, F., and Dezelee, S., Sigma virus: growth in *Drosophila melanogaster* cell culture; purification; protein composition and localization, *J. Gen. Virol.,* 65, 91, 1984.

64. Stollar, V., Togaviruses in cultured arthropod cells, in *The Togaviruses, Biology, Structure and Replication,* Schlesinger, R. W., Ed., Academic Press, New York, 1980, 583.

65. Richard-Molard, C. and Ohanessian, A., Méthode de clonage de cellules de Drosophile. Sensibilité aux rayons X de plusieurs lignées cellulaires, *Roux's Arch. Dev. Biol.,* 181, 135, 1977.

66. Richard-Molard, C., Etude du Cycle de Dévelopement de Rhabdovirus dans les Cellules d'Insecte en Utilisant des Cultures de Cellules de Drosophile, Thèse Paris-Orsay, 1984.

67. Contamine, D., The late functions of *Drosophila* sigma virus, *Arch. Virol.,* 82, 31, 1984.

68. Mahy, B. W. J., Minson, A. C., and Darby, G. K., *Virus Persistence,* the Society for General Microbiology, Symp. 33, Cambridge University Press, New York, 1982.

69. Eaton, B. T., Evidence for the synthesis of defective interfering particles by *Aedes albopictus* cells persistently infected with sindbis virus, *Virology,* 77, 843, 1977.

70. King, C. C., King, M. W., Garry, R. F., Wan, K. M. M., Ulug, E. T., and Waite, M. R. F., Effect of incubation time on the generation of defective-interfering particles during undiluted serial passage of sindbis virus in *Aedes albopictus* and chick cells, *Virology,* 96, 229, 1979.

71. Steacie, A. D. and Eaton, B. T., Properties of defective interfering particles of sindbis virus generated in vertebrate and mosquito cells, *J. Gen. Virol.,* 65, 333, 1984.

72. Laurent, J., Are the antigenic characteristics of vesicular stomatitis virus modified during evolution in *Drosophila?*, *Ann. Virol. Inst. Pasteur,* 134E, 465, 1983.

73. Enzmann, P. J., Induction of an interferon-like substance in persistently infected *Aedes albopictus* cells, *Arch. Gesamte Virusforsch.,* 41, 382, 1973.

74. Riedel, B. and Brown, D. T., Novel antiviral activity found in media of sindbis virus-persistently infected mosquito *(Aedes albopictus)* cell cultures, *J. Virol.,* 29, 51, 1979.

75. Newton, S. E. and Dalgarno, L., Antiviral activity released from *Aedes albopictus* cells persistently infected with semliki forest virus, *J. Virol.,* 47, 652, 1983.

76. Gillies, S. and Stollar, V., The production of high yields of infectious vesicular stomatitis virus in *A. albopictus* cells and comparisons with replication in BHK-21 cells, *Virology,* 107, 509, 1980.

77. Clinton, G. M. and Huang, A. S., Phosphoproteins of vesicular stomatitis virus and the host cell, in *Expression of Eukariotic Viral and Cellular Genes,* Petterson, R. F., Kääriäinen, L., Söderlund, H., and Oker-Blom, N., Eds., Academic Press, London, 1981, 163.

78. Blondel, D., Dezélée, S. and Wyers, F., Vesicular stomatitis virus growth in *Drosophila melanogaster* cells. II. Modifications of viral protein phosphorylation, *J. Gen. Virol.,* 64, 1793, 1983.

79. Newcomb, W. W. and Brown, J. C., Role of the vesicular stomatitis virus matrix protein in maintaining the viral nucleocapsid in the condensed form found in native virions, *J. Virol.,* 39, 295, 1981.

80. Reidler, J. A., Keller, P. M., Elson, E. L., and Lenard, J., A fluorescent photobleaching study of vesicular stomatitis virus infected BHK cells. Modulation of G protein mobility by M protein, *Biochemistry,* 20, 1345, 1981.

81. Rowlands, D., Grabau, E., Spindler, K., Jones, C., Semler, B., and Holland, J., Virus protein changes and RNA termini alterations evolving during persistent infection, *Cell,* 19, 871, 1980.

82. Wild, T. F. and Bijlenga, G., A rabies persistent infection in BHK21 cells, *J. Gen. Virol.,* 57, 169, 1981.

83. Weiss, B., Levis, R., and Schlesinger, S., Evolution of virus and defective interfering RNAs in BHK cells persistently infected with sindbis virus, *J. Virol.,* 48, 676, 1983.

84. Roux, L. and Waldvogel, F. A., Instability of the viral M protein in BHK 21 cells persistently infected with sendai virus, *Cell,* 28, 293, 1982.

85. Tuffereau, C., Lafay, F., and Flamand, A., Biochemical analysis of rabies virus proteins in three persistently infected BHK21 cell lines, *J. Gen. Virol.,* in press.

86. Jay, F. T., Dawood, M. R., and Friedman, R. M., Interferon induces the production of membrane protein-deficient and infectivity-defective vesicular stomatitis virions through interference in the virion assembly process, *J. Gen. Virol.,* 64, 707, 1983.

87. Thomas, P. S., Hybridization of denatured RNA and small DNA fragments transferred to nitrocellulose, *Proc. Natl. Acad. Sci. U.S.A.,* 77, 5201, 1980.

88. White, B. A. and Bancroft, F. C., Cytoplasmic dot hybridization. Simple analysis of relative mRNA levels in multiple small cell or tissue samples, *J. Biol. Chem.,* 257, 8569, 1982.

89. Stalder, J., Reigel, F., and Koblet, H., Defective viral RNAs in *Aedes albopictus* C6/36 cells persistently infected with semliki forest virus, *Virology,* 129, 247, 1983.

90. Leppert, M., Rittenhouse, L., Perrault, J., Summers, D. F., and Kolakofsky, D., Plus and minus strand leader RNAs in negative strand virus-infected cells, *Cell,* 18, 735, 1979.

91. Contamine, D., Two types of early mutants of *Drosophila* sigma virus, *Ann. Virol. Inst. Pasteur,* 131E, 113, 1980.

92. Flamand, A. and Lafay, F., Etude des mutants thermosensibles du virus de la stomatite vésiculaire appartenant au groupe de complémentation II. *Ann. Microbiol. Inst. Pasteur,* 124A, 261, 1973.

93. Coulon, P. and Contamine, D., Role of the *Drosophila* genome in sigma virus multiplication. II Host spectrum variants among the haP mutants, *Virology,* 123, 381, 1982.

94. Nakamura, N., Gay, P., and Contamine, D., Etude du locus *ref(2)P* de *Drosophila melanogaster.* I. Localisation cytogénétique de *ref(2)P, Biology of the Cell,* 56, 227, 1986.

95. Richard-Molard, C., Isolement de lignées cellulaires de *Drosophila melanogaster* de différents genotypes et étude de la multiplication de deux variants du rhabdovirus sigma dans ces lignées, *Arch. Virol.,* 47, 139, 1975.

96. Contamine, D., Role of the *Drosophila* genome in sigma virus multiplication. I. Role of the *ref(2)P* gene; selection of host-adapted mutants at the nonpermissive allele P*ᵖ*, *Virology,* 114, 474, 1981.

Index

INDEX

A

AA-20A cells, 27, 38, 39
Actinomycin D, 60, 117, 121, 123
Adsorptive endocytosis, 57
Adventitious agents, see also specific types
   in vaccine, 19
Aedes aegypti, 7, 16, 26, 38, 54, 69
Aedes albopictus
   alphaviruses in, 61, 93—94
   C6/36 cells from, 6—7
   cell lines from, 16
   cloned variants of, 69, 73, 117
   cytopathic effects and, 71, 93
   Dengue virus in, 49, 50
   DIP and, 119
   flaviviruses in, 46, 94—95
   monolayers of, 102
   persistent infection of, 68, 119
   pestivirus in, 94
   Semliki Forest virus in, 62, 71, 78, 93, 115
   Sindbis virus in, 70, 93, 102, 115, 119
   temperature-sensitive mutants in, 69
   togaviruses of, 93—94
   vesicular stomatitis virus of, 95
Aedes malayensis, 26
Aedes pseudoscutellaris, 6, 26, 46, 50, 78
   cytopathic effect and, 7
African green monkey kidney cells, 7
AG-55 cells, 28, 38, 39
Aggregation, 85
Alphaviruses, see also specific types, 46, 77—88
   Aedes albopictus of, 61, 93—94
   C-protein of, 71
   DI-particles of, 72
   genome of, 54, 57
   latent period for, 70
   mRNA of, 54
   nucleocapsid of, 59, 71
   proteins of, 62
   replication of, 61
   RNA of, 58
   RNA polymerase of, 57
   structural proteins of, 55
   togaviridae, 46
AM-60 cells, 27, 38—40
Amblyomma variegatum, 38
Amino acids, see also specific types, 92
Ammonium chloride, 84
Anopheles gambiae, 26, 38
Anopheles stephensi, 26
Anopheles A virus multiplication, 33
Antibodies, see also specific types
   flavivirus complex, 11
   monoclonal, see Monoclonal antibodies
Antiviral activity, 73
Antiviral substances, see also specific types, 69
AP-60 cells, 28

B

AP-61 cells
   cytopathic effects in, 40
   Dengue virus in, 49—51
   flavivirus in, 46
   kinetics of virus growth in, 38, 39
   minimum infectious dose in, 27
   monoclonal antibodies and, 6, 7
Arumowot virus, 47
AS-43 cells, 27, 38, 39
Assays, see also specific types
   plaque, 26
Assembly of nucleocapsids, 59, 114
Avidin-biotin, 40
Avidin D, 13

Banzi virus, 73
BHK cells, 26, 104
Biotin, 40
Biotinylated antimouse IgG, 13
BM-270 cells, 26, 38—40
Boophilus microplus, 26
Brain
   mosquito, 9
   suckling mouse (SMB), 26
Bromelain, 85
Budding, 60
Bunyaviridae, 47—48, 95
Bunyaviruses, 38, 47—48
Butandione, 86
Bwamba virus multiplication, 33
Bynyamwera virus multiplication, 33

C

C6/36 cells, 6, 46, 49, 51, 60, 95
   cloning of, 93
   cytopathic effects and, 79
   Dengue virus passage in, 16—19
   fusion and, 84
   safety tests on, 19
C-7 cells, 93, 95
C-7-10 cell cloning, 93
California encephalitis, 34, 38
L-Canavanine, 82
Capsid protein C, 57, 59
Capsids, 60
Carbobenzoxy-D-phenylalanyl-L-phenylalanyl-nitro-
   L-arginine, 84
Carbon dioxide sensitivity, 112
   in *Drosophila* sp., 117
Carboxylic acid metabolism, 48—49
Carriers, 112—113
CE, see California encephalitis
CEF, see Chick embryo fibroblasts
Cell fusing agent (CFA), 94

Cells, see also specific types, 26
  characteristics of, 44—45
  continuous, 128—129
  cured, 73, 119
  killing of, 95, 115
  mosquito, see Mosquito cells
  multilayered, 116
  persistently infected, 71—73
  transformation of, 68
  vertebrate vs. arthropod, 92, 102
  vertebrate vs. mosquito, 104—107
CF, see Complement fixation
CFA, see Cell fusing agent
Chandipura (CHP) virus, 37, 38
Chick embryo fibroblasts (CEF), 102, 103, 121
  Drosophila sp. vs., 123
CHIK, see Chikungunya virus
Chikungunya (CHIK) virus, 16, 38, 39, 46, 70, 71
  cytopathic effects and, 94
  multiplication of, 29
Chloroquine, 84
Chronic infections, defined, 68
Chymotrypsin, 82—84
Cleavages of proteins, 86
Cloned variants of *Aedes albopictus*, 69, 73, 117
Cloning, 93
Colcemide, 82
Complement fixation (CF), 4, 9, 12, 13
  Dengue virus and, 5
  yellow fever and, 9
Conformational change, 86
Contamination, 116
  Mycoplasma sp., 45
Continuous cells, 128—129
CPE, see Cytopathic effects
C-protein of alphaviruses, 71
Cross-reactivity, 5
Cured cells, 73, 119
Cycloheximide, 84
Cytochalasin B, 82
Cytolytic infections of mosquito cells, 69—70
Cytopathic effects (CPE), 40, 48, 78
  Aedes albopictus and, 71, 93
  Aedes pseudoscutellaris and, 7
  C6/36 cells and, 79
  chikungunya virus and, 94
  flavivirus and, 70
  Japanese encephalitis and, 94—95
  mosquito cells and, 61, 69, 92
  protein synthesis inhibition and, 97, 98
  Semliki Forest virus and, 94
  Sindbis virus and, 94
  temperature and, 70
  togavirus-infected mosquito cells and, 73
  vesicular stomatitis virus and, 95
  XTC-2 cells and, 26

D

Defective genomes, 114

Defective interfering particles (DIP), 61, 69, 73,
    116, 117
  Aedes albopictus and, 119
  in alphavirus, 72
  defined, 68
  Drosophila sp. and, 120
  in rhabdoviruses, 114
Defective interfering-RNA, 74
Defective RNA, 74
Dengue hemorrhagic fever, 8, 40
Dengue virus, 3—13, 40, 71, 73, 94
  complement fixation (CF) for, 5
  identification of, 4—5
  indirect immunofluorescence for, 11
  in mosquito brain tissue, 8
  multiplication of, 32
  passage of in C6/36 cells, 16—19
  plaque reduction neutralization test for, 4—5
  temperature sensitivity of, 17
  vaccines for, 15—20
  virologic surveillance for, 49—51
DFAT, see Direct immunofluorescent antibody test
DIP, see Defective interfering particles
Direct fluorescent antibody test (DFAT), 7, 11
Dissection of fusion process, 85—86
DNA polymerase, 68
DNA synthesis, 62
Dose, 27, 28
Drosophila sp., 111—130
  carbon dioxide sensitivity in, 117
  chicken embryo fibroblasts vs., 123
  DIP and, 120
  genes of, 127, 128
  mRNA in, 122
  Piry virus in, 126
  refractory genes of, 127
  rhabdoviruses in, 111, 115—119
  sigma virus in, 126, 128
  VSV leader RNA in, 125—126
  VSV transcription in, 124—125
DTT, 86

E

Eastern equine encephalitis (EEE), 46
EEE, see Eastern equine encephalitis
Electron microscopy, 19, 55, 59, 116
  of Sindbis virus, 104
Electrophoresis in polyacrylamide gels, see Poly-
    acrylamide gel electrophoresis (PAGE)
Encephalitis, see also specific types
  Eastern equine (EEE), 46
  St. Louis (SLE), 47
  Western equine (WEE), 69
Endocytosis, 57
Endoplasmatic reticulum, 59, 60
Envelope
  formation of, 59—60
  glycoproteins of, 104
  proteins of, 56, 62

Enzymes, see also specific types
  malic, 44
  proteolytic, 82, 85
Evolution of genome, 16

F

Fibroblasts
  chick embryo, see Chick embryo fibroblasts
    (CEF)
  rhesus monkey lung, 17
Flaviridae, 46—47
Flavivirus sp., 9, 73
Flavivirus complex antibody, 11
Flaviviruses, see also specific types, 46—47, 60,
    62, 71
  Aedes albopictus, 94—95
  cytopathic effects and, 70
  morphogenesis of, 60
  morphology of, 55
  nucleocapsids of, 60
  replication of, 70
  RNA of, 54—55
  RNA replication in, 58—59
Fluorescein-labeled Avidin D, 13
Fluorescent antibody test, see Immunofluorescence
p-Fluorophenylalanine, 82
Fusion, 79—80, 84—85
  C6/36 cells and, 84
  dissection of process of, 85—86
  duration of, 80
  inhibitors of, 80, 83
  morphology of, 80
  patchwise, 80
  triggering of, 85—86
Fusogenic factor, 84—85

G

Ganjam virus, 26, 35
Genes, see also specific types
  Drosophila, 127, 128
  refractory, 113—115, 127
Genome probe, 124
Genomes
  alphaviral, 54, 57
  defective, 114
  evolution of, 16
  togavirus, 54—55
Germiston virus, 34, 95
Getah virus, 46
Glycoproteins, see also specific types, 121
  envelope, 104
  transmembranous, 60
Growth
  kinetics of, 27, 38—39
  of rhabdoviruses, 117—119
  of togaviruses, 56

H

Hemagglutination-inhibition (HI) and yellow fever,
    9
Hereditary transmission of sigma virus, 119
Heterogeneity of virus population, 128—129
HIMAF, see Hyperimmune mouse ascitic fluid
Host-cell DNA synthesis, 62
Host cell macromolecular synthesis, 62—63
Host-cell mRNA translation, 63
Host-cell RNA synthesis, 62
Host protein synthesis, 71, 72, 125
Host-range mutants of Sindbis virus, 102—107
Hydroxy acid metabolism, 48—49
Hyperimmune mouse ascitic fluid (HIMAF), 4, 5,
    7, 8, 11
Hypersensitivity to vaccine, 19

I

Identification of Dengue virus, 4—5
IFAT, see Indirect immunofluorescence
Immunocytochemistry, 40
Immunofluorescence, 7
  direct (DFAT), 7, 11
  indirect, see Indirect immunofluorescence
  of rhabdovirus, 115, 117
Immunoglobulin G, 13
Immunoperoxidase, 40
Indirect immunofluorescence
  for Dengue viruses, 11, 40
  monoclonal antibodies and, 5—9
Infections, see also specific types
  chronic, 68
  conditions of, 116—117
  latent, 68
  multiplicities of (MOI), 78, 79, 85
  persistent, see Persistent infections
  slow, 68
  time course of, 79—80
Infectious dose, 27, 28
Inhibition, see also specific types
  of cellular RNA transcription, 125
  of fusion, 80, 83
  of protease, 84, 86
  of protein synthesis, 96—99
Inner-membrane M protein of VSV, 122
Insect cells, see also specific types
  persistent infection in, 119—126
Insect RNA polymerases, 125
Interference phenomenon, 74
Interferon, 69, 124
Interferon-like substances, see also specific types,
    73, 121
Interpretation of persistence, 73—75
Ionophores, 85
Isocitrate dehydrogenase, 44
Isoenzyme analyses of vaccine, 19
Isolation
  monoclonal antibodies in, 6—9

of mosquito cells, 44
Isozymes, see also specific types, 44

J

Japanese encephalitis (JE), 38—40, 47
  cytopathic effects and, 94—95
  multiplication of, 31
JE, see Japanese encephalitis

K

Karimabad virus, 47
Karyology, 44
  of vaccine, 19
Keterah virus multiplication, 36
Killing of cells, 95, 115
Kinetics
  of arbovirus growth in cell cultures, 27, 38—39
  of persistence, 70—71
  of rhabdovirus production, 117
Kunjin virus, 71

L

Labeling, see Radiolabeled
Lactalbumin hydrolysate, 92
Langat (LGT) virus, 35, 38—40
Latent infections, defined, 68
Latent period of alphaviruses, 70
Leader RNA, 125—126
Leupeptin, 84
Light microscopy, 80
Lipids, see also specific types, 62
LLC-MK$_2$ cells, 4, 7—8, 11, 48, 50
Louping ill (LI) virus, 35, 38, 40
LSTM-AA-20A cells, 26
LSTM-AG-55 cells, 26
LSTM-AM-60 cells, 26
LSTM-AP-61 cells, 26
LSTM-AS-43 cells, 26
LSTM-BM-270 cells, 26
Lysosomotropic agents, 84
Lyssaviruses, 112

M

Macromolecular synthesis, 62—63
Malic enzymes, 44
Mammalian cell cultures, 7—8
Mayaro (MAY) virus, 30, 38
Membrane lipids, 62
Metabolism, 48—49
Methionine, 99
Middleburg virus, 26, 30
Minimum infectious dose, 27, 28
Mitochondrial RNA (mRNA)
  alphavirus, 54

Drosophila sp., 122
  modification of, 54
  processing of, 54
  translation of host-cell, 63
Modification of mRNA, 54
MOI, see Multiplicities of infection
Molecular biology, 54—55
Monensin, 85
Monoclonal antibodies, 3—13, 40
  with isolation systems, 6—9
  sensitivity of, 13
  serotype-specific, 5
  specificity of, 9—12
Monolayers, 48, 79, 85, 86
  Aedes albopictus, 102
Morphogenesis of flaviviruses, 60
Morphology, 44
  of flaviviruses, 55
  of fusion, 80
Mosquito cells
  alphavirus replication in, 61
  cytolytic infections of, 69—70
  cytopathic effect and, 61, 69, 92
  isolation of, 44
  persistent infections of by togaviruses, 67
  Sindbis virus mutants in, 102—104
  Sindbis virus replication in, 104—107
  togavirus-infected, 73—75
  vertebrate cells vs., 104—107
Mosquito tissue, 8—9
M protein, 122, 123
mRNA, see Mitochondrial RNA
Multilayered cells, 116
Multiplicities of infection (MOI), 78, 79, 85
Mutants, see also specific types
  Piry virus, 116
  plaque, see Plaque variants
  sigma virus, 116, 127
  Sindbis virus, 102—107
  small plaque, 69
  temperature-sensitive, 61, 69, 74, 113, 119, 127
Mycoplasma sp., 45

N

Ndumu virus multiplication, 30
Neutralization, 4—5, 48
Newcastle disease virus, 84
Nucleic acid, 68
  synthesis of, 114
Nucleocapsids
  alphavirus, 59, 71
  assembly of, 59, 114
  flavivirus, 60

O

Oligosaccharides, 106
O'Nyong-Nyong (ONN) virus, 31, 38

Orbivirus, 48
Organization of virion, 55
Outer membrane protein, 121

P

Pactamycin, 99
Pacui virus, 47
PAGE, see Polyacrylamide gel electrophoresis
Patchwise fusion, 80
Persistence, 115
  interpretation of, 73—75
  kinetics of, 70—71
  mechanisms of, 68—69
Persistent infections, 39—40, 48
  Aedes albopictus, 119
  defined, 68
  in insect cells, 119
  togavirus, 67
  in vertebrate cells, 119—121
Persistently infected cells, 71—73
Pestivirus in *Aedes albopictus*, 94
pH
  changes in, 79
  temperature and, 80
Phenylmethylsulfonylfluoride, 84
Phlebovirus, 28, 47—48
Phosphoglucoisomerase, 44
Phosphoglucomutase, 44
Phospholipids, see also specific types
  of Semliki Forest virus, 62
Phosphoproteins, see also specific types, 97
Physical properties, see also specific properties
  of togavirus genome, 54—55
Piry virus, 37, 113, 115—117
  Drosophila refractory genes and, 127
  Drosophila sp. and, 126
  heterogeneity of, 128—129
Plaque assay, 26
Plaque development, 48
Plaque reduction neutralization test (PRNT), 9, 12,
    13, 48
  for Dengue virus, 4—5
Plaque variants, 39, 61, 69, 72
Podophyllotoxin, 82
Poliovirus, 102
Polyacrylamide gel electrophoresis (PAGE), 54, 71,
    84, 86
  of pestivirus, 94
  of VSV, 125
PR-6 cells, 46
PRNT, see Plaque reduction neutralization test
Production of viruses, 79
Protease inhibitors, 84, 86
Proteinase K, 84, 85
Protein G deficiency, 123
Proteins, see also specific types, 80
  in alphavirus, 62
  cleavages of, 86
  envelope, 62
    inhibition of, 96
    outer membrane, 121
    in rabies, 123
    in Sindbis virus, 123
    spike, 78
    structural, 55, 107
    transmembrane, 86
    viral envelope, 56
    in VSV, 123
Protein synthesis, 57—58, 71—72
  host, 71, 72, 97
  inhibition of, 96—99
  VSV, 121—124
Proteolytic enzymes, 82, 85
PS cells, 26
Punto Toro virus, 47
Putative fusogenic factor, 84—85

Q

Quaranfil (QRF) virus, 36, 38

R

RA-243 cells, 40
Rabies, 112, 123
Radiolabeled genome probe, 124
Radiolabeled methionine, 99
Radiolabeled M protein, 123
Radiolabeled phosphoproteins, 97
Radiolabeled uridine, 102
Receptors, see also specific types
  for Semliki Forest virus, 57
Refractory gene sensitivity, 113—115
Replication, 27, 45—49
  of alphaviruses, 61
  early events in, 56—57
  of flaviviruses, 70
  of flavivirus RNA, 58—59
  host cell macromolecular synthesis and, 62—63
  of Semliki Forest virus, 81
Retroviruses, see also specific types, 19, 68
Reverse transcriptase, 19, 68
Rhabdoviruses, see also specific types, 38, 95,
    111—130
  defective interfering particles in, 114
  Drosophila sp., 111, 115—119
  growth cycles of, 117—119
  production kinetics of, 117
Rhesus monkey lung fibroblasts, 17
Rhipicephalus appendiculatus, 40
Ribonucleoprotein (RNP), 113
RML-12 cells, 7
RNA
  alphavirus, 58
  defective, 74
  flavivirus, 54—55, 58—59
  inhibition of cellular transcription of, 125
  mitochondrial, see Mitochondrial RNA (mRNA)

Sindbis virus, 54
VSV leader, 115
RNA polymerase, 113, 114
  alphavirus, 57
  insect, 125
  Sindbis virus, 102
  VSV, 115
RNA replicase, 58
RNA synthesis, 58—59, 70, 72, 96
  host-cell, 62
  Semliki Forest virus and, 125
  VSV, 124—125
RNP, see Ribonucleoprotein
Ross River virus, 46, 70
Rubella vaccine, 16

S

Safety tests on C6/36 cells, 19
Sandflies, 113
Sandfly fever virus, 27, 34, 38, 47
Self-curing, 71, 73, 74, 119
Semliki Forest virus, 38—40, 70, 119
  in *Aedes albopictus*, 71, 78, 93, 115
  cell death and, 95
  cytopathic effects and, 94
  multiplication of, 29
  phospholipids of, 62
  receptors for, 57
  replication of, 81
  RNA synthesis and, 125
Sendai virus M protein, 123
Sensitivity
  to carbon dioxide, 112, 117
  of monoclonal antibodies, 13
  to refractory genes, 113—115
Serotype-specific monoclonal antibodies, 5
SFV, see Semliki Forest virus
Shutoff phenomenon, 63
Sialic acid, 62, 104, 106
Sigma virus, 112, 113, 115, 117, 118
  cycle of in flies, 126—127
  Drosophila sp. and, 126, 128
  hereditary transmission of, 119
  mutants of, 116
  temperature-sensitive mutants of, 127
Sindbis virus
  in *Aedes albopictus*, 70, 93, 102, 115, 119
  cell death and, 95
  cytopathic effects and, 93, 94
  electron microscopy of, 104
  fusion of, 86
  glycoproteins of, 55
  host-range mutants of, 102—107
  in mosquito cells, 46
  multiplication of, 29
  persistent infections of, 48, 68, 69, 71—73
  proteins in, 123
  protein synthesis in, 98
  replication of in mosquito cells, 104—107

restriction of, 106
RNA of, 54
RNA polymerase of, 102
RNA synthesiš in, 58
small plaque variants of, 39
Skin tests for vaccine, 19
SLE, see St. Louis encephalitis
Slow infections, defined, 68
Small plaque variants, 39, 61, 69
  temperature-sensitive, 72
SMB, see Suckling mouse brain
Specificity of monoclonal antibodies, 9—12
Spike proteins, 78
Spikes, 85
Spinner cultures, 85
St. Louis encephalitis (SLE), 47
Structural proteins, 55, 107
Suckling mouse brain (SMB), 26
Surveillance for dengue virus, 49—51
Syncytia, 79, 80, 85, 86, 94
  formation of, 61
Synthesis
  DNA, 62
  host cell macromolecular, 62—63
  host RNA, 125
  nucleic acid, 114
  protein, see Protein synthesis
  RNA, see RNA synthesis
  of viral components in *Drosophila* sp., 121
  of viral components in vertebrate cells, 121

T

TA-9 cells, 19, 44, 46—48
TA-42 cells, 19, 44, 46—48
TBE, see Tick-borne encephalitis
TCA, 97, 99
Temperature
  cytopathic effects and, 70
  Dengue virus and, 17
  pH and, 80
Temperature-sensitive mutants, 61, 69, 72—74,
    113, 119
  of sigma virus, 127
Temperature-sensitivity of VSV, 116
Tetrathionate, 86
Thermolysin, 84
Tick-borne encephalitis (TBE), 68
Time course of infection, 79—80
TNBS, see Trinitrobenzenesulfonic acid
Togaviruses, see also specific types, 68
  Aedes albopictus of, 93—94
  genome of, 54—55
  growth cycle of, 56
  molecular biology of, 54
  in mosquito cells, 73—75
  organization of, 55
  persistent infections of mosquito cells by, 67
p-Tosyl-L-argininemethylester, 84
N-α-Tosyl-L-lysyl-chloromethylketone, 84

N-α-Tosyl-L-phenylalanyl-chloromethylketone
(TPCK), 86
Toxorhynchites amboinensis, 6, 7, 16, 19, 44
TPCK, see N-α-Tosyl-L-phenylalanyl-
chloromethylketone
TRA-171 cells, 44, 46—50
TRA-272 cells, 44
TRA-284 cells, 6, 7, 19, 45, 50
TRA-284-SF cells, 11, 48, 49, 51
Transcriptase, 68
Transcription of cellular RNA, 125
Transformation of cells, 68
Translation of host-cell mRNA, 63
Transmembrane glycoproteins, 60
Transmembrane proteins, 86
Triggering of fusion, 85—86
Trinitrobenzenesulfonic acid (TNBS), 86
Trypsin, 82
Ts, see Temperature-sensitive
Tumorigenicity of vaccine, 19
Tumor viruses, 68
Tunicamycin, 84, 104

U

Uncloned cell populations, 93
Uridine, 102

V

Vaccines, see also specific types
adventitious agents in, 19
for Dengue viruses, 15—20
hypersensitivity to, 19
isoenzyme analyses of, 19
karyologic analyses of, 19
for rubella, 16
skin tests for, 19
tumorigenicity of, 19
for yellow fever, 16
Vero cells, 7, 8, 26
Vertebrate cells
alphavirus replication in, 61
arthropod cells vs., 60—62, 92, 102
mosquito cells vs., 104—107
persistent infection in, 119—121
restriction of Sindbis virus in, 106
synthesis of viral components in, 121
VSV RNA synthesis and, 125
Vesicular stomatitis virus (VSV), 112, 117, 119
in *Aedes albopictus*, 95—102

cell killing and, 115
inner-membrane M protein of, 122
interferon and, 124
PAGE of, 125
proteins in, 123
RNA polymerases in, 115
sandflies and, 113
thermosensitivity of, 116
Vesicular stomatitis virus (VSV) leader RNA, 115,
125—126
Vesicular stomatitis virus (VSV) protein synthesis,
121—124
Vesicular stomatitis virus (VSV) RNA synthesis,
124—126
Vesicular stomatitis virus (VSV) transcription in
Drosophila sp., 124—125
Vesiculoviruses, see also specific types, 47, 112,
113
Vinblastine, 82
Vincristine, 82
Viral membrane lipids, 62
Vitamins, see also specific types, 92
VSV, see Vesicular stomatitis virus

W

WEE, see Western equine encephalitis
Western equine encephalitis (WEE), 69
West Nile (WN) virus, 32, 38—40, 47, 94
WN, see West Nile virus

X

Xenopus laevis, 26
XTC-2 cells, 26, 38, 40

Y

Yeastolate, 92
Yellow fever, 38, 40, 47
complement fixation and, 9
HI and, 9
multiplication of, 32
vaccine for, 16

Z

Zika virus, 31, 38
Zirqa virus, 36, 38